PRODU ... ON

Materials Technology

VOLUME 1

Materials Technology

VOLUME 1

THE NATURE OF MATERIALS

J. G. TWEEDDALE
F.I.M., F.Weld.I., C.Eng.,
M.I.Mech.E.

Senior lecturer (*Fabrication Metallurgy*)
Imperial College of Science and Technology, London

LONDON
BUTTERWORTHS

THE BUTTERWORTH GROUP

ENGLAND
Butterworth & Co (Publishers) Ltd
London: 88 Kingsway, WC2B 6AB

AUSTRALIA
Butterworths Pty Ltd
Sydney: 586 Pacific Highway, NSW 2067
Melbourne: 343 Little Collins Street, 3000
Brisbane: 240 Queen Street, 4000

CANADA
Butterworth & Co (Canada) Ltd
Toronto: 14 Curity Avenue, 374

NEW ZEALAND
Butterworths of New Zealand Ltd
Wellington: 26–28 Waring Taylor Street, 1

SOUTH AFRICA
Butterworth & Co (South Africa) (Pty) Ltd
Durban: 152–154 Gale Street

First published 1973

ISBN 0 408 70391 1 Standard
0 408 70392 X Limp

Printed in Great Britain by
Redwood Press Limited,
Trowbridge, Wiltshire.

Preface

Materials technology is now universally recognised as the basis upon which all engineering skills and crafts are founded. As a result, every engineering craft, technical and technological training course incorporates studies related to materials technology. Some courses require an intensive understanding and others a more cursory knowledge, but there is a lack of reliable, simple, comprehensive textbooks which can give a usefully broad introduction to those who are about to enter on an intensive course and a reasonably detailed survey for those with a less specialised interest.

This work is offered in the belief that it may go some way towards meeting this need. The theme has been separated in what seems a logical division of two volumes, one dealing with the nature of materials and the other with systems of treatment. Neither volume is really complete without the other, but for many courses of study it is unlikely that both will have to be purchased at one time.

The volumes are intended to cover the whole materials content of those City and Guilds of London Institute engineering technician courses and those Higher National Diploma courses which do not require specialised materials study. They should also serve as useful broad surveys leading into more specialised technician courses and as introductory texts for first year university undergraduates who are meeting materials for the first time.

The present volume starts with the atom and leads on to consider how the energies of interaction within and between atoms are controlled, and what are the ideal and real natures of physical and chemical properties. The influence of defects is also discussed and how knowledge of the mechanism of defects leads to the development of new materials.

My sincere gratitude is extended to all those friends, too numerous to mention by name, who have helped by discussion and comment to clarify my ideas. Needless to say any mistakes are my own responsibility.

J. G. Tweeddale

Imperial College 1972

Preface

Contents

1 INTRODUCTION 1

2 THE NATURE OF MATERIALS 12

3 THE MAKEUP OF MATERIALS 38

4 BASIC AND SIGNIFICANT PROPERTIES 70

5 IDEAL AND REAL PROPERTIES 105

6 CONTROL OF USEFUL PROPERTIES 147

INDEX 201

1 Introduction

In the study of evolving civilisation it is customary to divide the development into ages identified by what is the major development characteristic of that age. Thus we have a Stone Age, a Bronze Age and an Iron Age leading towards the present age, often identified as an Age of Technology. Each title, except the last, is a direct reference to the most advanced type of constructional material used. The last title refers partly to the complexity of tools and weapons and partly to the materials because it is no longer possible to pick out one material that alone characterises the age. The word *material* basically means anything that has a characteristic identity recognisable by the two senses of sight and touch, but today it is more commonly used as a shortening of *material of construction* or *material of manufacture*.

As modern artefacts have developed in complexity and range from simple kitchen utensils to computer-operated factories, so have modern constructional materials. Anyone desiring to get the best use, or even simple enjoyment out of the amenities of modern civilisation must take an active interest in materials. The housewife, whether she does it consciously or instinctively, takes interest in a variety of ways. She is concerned with the strength, appearance (perhaps particularly with the ease of maintenance of appearance) and the *feel* of each kind of domestic tool that she may use. She is likely to be interested in the aesthetic appearance and durability of materials used purely for pleasure or decoration such as jewellery or ornaments. On the other hand, there are individuals who are more particularly associated with the engineering design, manufacture and operation of equipment who must be directly interested in the materials used.

This wide range of interest makes a knowledge of the technology of materials a highly desirable adjunct to all education, and for a practising technologist, an essential branch of knowledge.

The problem in acquiring such knowledge is that so many branches of learning are incorporated into materials technology that it becomes difficult to comprehend without the benefit of a thorough grounding in the pure and applied sciences. Hence, for the broad approach intended in this book, it is inevitable that the treatment of many individual topics can only be descriptive. However this can be an advantage since for the less specialised reader, a descriptive approach provides a bridge over the fact that much of the detailed theory is complex and full of alternative lines of argument.

The remainder of this chapter is devoted to a consideration of the scope of materials technology so that the various branches of the subject can be seen in perspective.

1.1 THE RANGE OF MATERIALS TECHNOLOGY

Each part of materials technology is based on the application of scientific principles of one sort or another and the range of sciences that can be involved embraces biology, chemistry, economics, geology, mathematics, physics, psychology and sociology. Of course all these disciplines do not apply simultaneously in every situation, but usually several apply together. Probably the only ones that apply without exception are economics and sociology because a material is neither made nor used if no one wants it sufficiently to pay for it,

The application of these sciences resolves itself into two main complementary fields of interest: (a) properties of materials and (b) processes of manufacture.

1.2 PROPERTIES OF MATERIALS

To justify the incorporation of a particular material into any artefact it must have at least one desirable property. A good example of this is lubricating oil with its facility for reducing rubbing friction. However, one desirable property alone would rarely qualify a material for use. Even in the simple case of appearance, that property alone would rarely be sufficient no matter how well it looked. It would also have to have some degree of durability and it would have to be possible to prepare the material in a form suitable for locating and holding in position to fulfil its primary function. The more numerous the useful properties possessed by a material, the more likely it is to be used in differing applications.

Relevant to nearly every material of construction are at least four principal groupings of properties: (1) controllability of structure (2) formability (3) physical properties and (4) chemical properties. The importance of each group will vary with each type of application.

CONTROLLABILITY OF STRUCTURE means the facility with which the *internal* state of the material can be controlled to a condition of uniform consistency in a given component so that the properties are predictably, uniform and consistently reliable in service.

FORMABILITY is the capacity of a material (once it is in a suitably controlled state of internal structure) to withstand the different types of forming operation that may have to be applied to produce a particular shape of component. It applies only to materials used in the solid state, but this means the majority of constructional materials.

PHYSICAL PROPERTIES include all properties such as mechanical strength, electromagnetic properties, thermal properties and optical properties.

CHEMICAL PROPERTIES include all those properties related to interaction between a material's internal structure and its service environment. These interactions include effects such as corrosion, which can greatly influence the stability of a material in service.

The four main groupings are each made up of patterns of intricate inter-relationships between the specific properties and these relationships depend to a large extent on particular circumstances. In fact it might well be that one specific property could be important to more than one of the groups (e.g. tensile strength in relation to formability and to applied physical properties). Some of these relationships are dealt with in more detail in later chapters. Each group is considered briefly in a little more detail in later parts of this section.

No matter how unspecific these groups tend to appear, one thing must be stated quite definitely, namely, that every one of the specific properties that go to make up a group property derives from the atomic nature of material.

1.3 ATOMIC NATURE OF MATERIAL

Fundamental forms of energy appear to interact with each other according to very consistent mathematical rules which make stable energy associations fall into well defined patterns. Certain of these association patterns make up the characteristic basic bricks of matter known as *atoms*. Atoms are particular energy groupings which persist (i.e. are stable) long enough for them to appear to be possessed of an infinite life if they are not reacted on by relatively large or abnormal external forces. There are about 100 types of these more stable atoms existing in various forms and associations.

Atoms can interact with each other in different ways according to their make up, by means of the energy links or *bonds* that form between them. Several different kinds of bonds are possible, sometimes in association with each other and sometimes almost in isolation. When such bonds form between a sufficient number of atoms, a material is formed with properties governed partly by the kind of links that prevail and partly by the nature of the constituent atoms. Thus a material (used in the broadest sense of the word) can be in a state typical of a gas (very loosely linked), a liquid (closely but not rigidly linked), or a solid (closely and rigidly linked) according to the type of connection between the constituent atoms in the prevailing ambient conditions. Such a material will possess properties characteristic of the types of atom that go to make it up and of the system of bonding.

Developing, controlling and ordering the type of bonding feasible in a given material makes it possible to change or develop some properties of that material within certain limits. This is the basis upon which materials technology is founded.

1.4 STRUCTURAL CONTROLLABILITY

One aspect of the study of the properties of material is the control of the positioning and bonding of the atoms within a solid or liquid material, since any change in the basic internal features may change many of its potentially important properties.

In achieving this control there are two main problems to be overcome. (1) The whole internal structure of a mass of material for practical use must be uniformly ordered to ensure equally predictable behaviour for each part of the mass that is used. (2) Particular desirable properties must be so effectively controlled and developed that the material fulfils, for the duration of its intended life, the purpose for which it is intended.

In practice these problems are not always easy to solve. Most materials have to be prepared from existing natural materials and it is rare for these natural materials to be either in a pure condition or in a suitable internal structural state for immediate use. In nearly every case the natural materials have to be radicaliy changed before they are usable. Impurities, accidentally or normally present, have to be removed (often requiring expensive multistage processes) and desirable compositional additions may have to be made. Even after this, further treatments may be required to change the internal order of the constituent atoms. Preparations of these kinds can be expensive and many highly desirable materials are unlikely ever to be used, simply because they are too costly to produce.

Structural controllability can subdivide into a number of specific aspects namely, crystalline, intercrystalline, molecular and intermolecular. These terms refer to the type of atom structure existing within a material (and more particularly within a solid material). The actual phenomena will be dealt with in the relevant chapters since at this stage we are concerned only with their places in the general picture.

The atoms that make up a material can associate with each other in certain specific ways. In a pure solid made up of only one kind of atoms the commonest arrangement is *crystalline* (see Chapter 3) in which the very large numbers of constituent atoms are systematically arranged relative to each other in a repetitive, three dimensional geometric pattern. It is rare for the whole of the mass of such a material to be made up from one continuous orientation of pattern, and so a number of distinguishable zones, usually of differing orientation, known as *grains* are likely, each zone being separated by an *intercrystalline boundary* or *grain boundary*. Control of (1) the type of atomic pattern; (2) the size of the grains, and (3) the type of boundary arrangement, can each be used to influence the properties of a material in which a crystalline structure is present. Many materials made up from more than one kind of atom may also solidify into similar crystalline types of pattern, perhaps with differing grains having differing atomic array patterns and with varied intercrystalline properties.

In many materials, particularly those containing more than one kind of atom, the internal arrangement may be very different, with small equal numbers of atoms (and fixed numbers of atom types where relevant) forming themselves into individual stable geometric groups called *molecules*. Large numbers of molecules linked to each other by varied types of inter-molecular forces, acting across the *molecular boundaries* (also called grain boundaries in some situations) can group to form the solid or the liquid state of the material. Here again, by changing (1) the atomic array within the molecules, (2) the type of molecular pattern and (3) the linking between the molecules, the properties of the material can be changed and controlled.

Some materials may include crystalline and molecular states simultaneously within themselves or be capable of being changed from one state to the other. Many people regard a crystal as a special form of molecule but the differences between their many features are sufficiently marked for us to regard them as separate phenomena.

1.5 FORMABILITY

Although it is essential to be able to control the internal structure of a solid material, this kind of control is not the only criterion of usefulness if the material is to be of practical use. It is equally important to be able to change a mass of the solid material into predetermined shapes with dimensional accuracy suited to the intended applications. Formability subdivides into four closely related sub branches each one referring to the capacity of the material to undergo a particular mode of forming. These four capacities are castability, machinability, plasticity and weldability.

CASTABILITY is a measure of the ability of a material to withstand the operation of solidifying from a liquid into a predetermined solid shape whilst achieving effective continuity of the internal structure of the shaped mass. It also involves consideration of means for preparing the material in a suitable liquid form. The latter is usually a simple matter of controlled melting of a fusible material by application of heat, but it could involve the preparation of basic raw materials in the form of a suitable liquid or slurry mixture which is subsequently irreversibly solidified, by chemical reaction, into the required solid material and shape (e.g. concrete).

MACHINABILITY is a measure of the ability of a solid material to withstand the process of shaping by the forcible removal of undesired portions of the initial solid mass. There are several forms of machining including (1) mechanical removal of *chips* (2) burning away of surplus material (thermal cutting) (3) chemical dissolving of unwanted material (4) electrolytic dissolution of unwanted material and (5) electric spark erosion of unwanted material. The mechanical machining methods are probably the best known of all forming processes but they are not by any means the most important.

PLASTICITY is a measure of the ability of a material to have its shape permanently changed by force without fracture or damage to the internal structure. There are many complex forcible forming processes but most of them are made up from combinations of three simple basic operations (1) bending (2) spreading and (3) attenuation by stretching or extruding.

WELDABILITY is the measure of the ability of a material to withstand the application of integral bonding operations to join one part of a material to another part of the same or a different material so that the final product forms a material continuum. These operations make use of the atomic bonding characteristics of the materials making up the welded assembly and are likely to be most readily applicable to fusible materials. Castability and plasticity are usually closely linked with weldability.

An understanding of each of these aspects of formability is essential if the factors governing the processes of manufacture are to be understood.

1.6 PROPERTIES OF APPLICATION

Properties of application may seem to be a rather cumbersome title but, since it is normal practice in materials literature to classify materials under the heading of those properties required from them in service as against those required for manufacture, the distinctive title is desirable.

As mentioned in Section 1.2, properties of application are normally grouped into two main divisions, physical and chemical, and the two main groups subdivide into more specialised subgroups. Thus the physical properties include appearance, electrical, fluid, magnetic, mechanical, optical and thermal properties whilst chemical properties subdivide into corrosion resistance, oxidation resistance and reactivity. The majority of materials are used because they possess one or more of the physical properties in outstanding degree. Although the chemical properties of a material may seem to be unimportant in applications other than particular specialised ones, it must not be forgotten that in almost every application, corrosion resistance will be a particular factor.

It is desirable, for even the most elementary understanding of materials, to recognise the relative significance of each of these various subdivisions even although they are not all always of equal importance. In the following paragraphs the scope of each subdivision is outlined.

ELECTRICAL PROPERTIES such as electrical conductivity (the measure of freedom for passage of an electric current through a material) or, conversely, resistivity (the measure of the opposition to free passage of electric current through a material) are of obvious importance in electrical equipment. There are also those effects of alternating current flow in particular cases which can give rise to internal heating effects in both conductive materials (by electromagnetic induction) and in nonconductive materials (by dielectric heating). Magnetic properties are closely linked with electrical properties so electromagnetic properties can be of particular importance in many types of construction situation.

FLUID PROPERTIES are significant in all those normally liquid materials which are used as lubricating, power transmission or heat transfer media. The freedom with which a liquid flows in particular situations directly influences the power which it absorbs in the act of flowing. In some situations a high power loss is desirable (e.g. in a fluid shock absorber). In others (e.g. lubrication and power transmission) a low power loss is desirable. Fluid flow of transiently molten materials is important in casting operations. Also certain normally solid but plastic materials are used as very viscous fluids.

MAGNETIC PROPERTIES can be of three kinds namely *diamagnetic* (repelled by a magnetic field) *ferromagnetic* (strongly attracted by a magnetic field), *paramagnetic* (weakly attracted by a magnetic field). *Diamagnetism* and *paramagnetism* are very weak forces in all those known materials in which they occur, so they are generally disregarded as properties of application. Ferromagnetic properties are specifically important in many applications such as in instrumentation, in the use of electric power and in some industrial treatment processes but are often of incidental importance because of the need to control stray magnetic and electromagnetic effects which may harm the function of a mechanism.

MECHANICAL PROPERTIES are the most obviously important properties of application since mechanical strength affects almost every application. Because of this, these properties are often regarded from a completely distorted aspect, sometimes to the exclusion of other important properties. Within themselves mechanical properties include not only the simple tensile compressive and shearing strengths (resistance to fracture by pulling, pushing and internal sliding respectively) but also the resistances shown to different forms of elastic and plastic deformation, geometric discontinuities (notches), repeated loading (fatigue strength), shock loading (toughness) and vibrations. The importance of these other properties is not so clearly recognised because they are often difficult to assess.

OPTICAL PROPERTIES are significant only to a small number of materials, notably those which are either sufficiently transparent to be used for light transmission or can be made sufficiently light reflective to serve usefully as reflectors. Because of their specialised nature optical properties will not be considered further.

THERMAL PROPERTIES such as conductivity, specific heat, thermal expansion and contraction, and structural change phenomena (notably solidification, melting and boiling) can be important in materials intended for applications involving heat transfer, heat insulation, fluctuating temperature or elevated temperature. The relative importance of each property will depend on the particular type of service and usually the thermal properties will have to be associated with one or more other properties such as mechanical strength, corrosion resistance or refractoriness.

CORROSION RESISTANCE PROPERTIES are the properties of resistance of a material to chemical solution or disintegration of its structure in differing service conditions. Corrosion properties differ relative to environments and there are certain applications in which a particular type of corrosion resistance is the principal property, notably in linings for vessels containing corrosive substances. Normally corrosion resistance is a property requirement incidental to other requirements in a material.

REACTIVITY is the inverse of corrosion resistance in the sense that it is a measure of the tendencies of a material to react with other substances. It is particularly important in an energy generating material such as fuel or in a chemically reacting material used to produce a desirable reaction product or in a material used as a catalyst where the presence of the material speeds up a reaction without itself taking part.

REFRACTORINESS is the resistance of a material to change or disintegration at elevated temperature. It is related both to a particular form of direct corrosion attack by environmental oxygen (oxidation) and to the melting properties of the material each of which can influence behaviour. It is of major importance in such things as combustion chamber linings, rocket motor casings and similar applications.

1.7 PROCESSES OF MANUFACTURE

Although desirable properties of application and even a reasonable degree of formability may be inherent in a material, it may yet be difficult to make reasonable use of that material if the technology for manufacturing it into the finished form is misunderstood or undeveloped. Therefore a critically important part of the materials technology field is the technology of the processes of manufacture.

A very important branch of process technology is the study of possible applications of processes to the manufacture of materials used and potentially usable in the differing fields of construction. Then there is a large branch concerned with the actual production of components and assemblies, covering planned control of the quality of materials and products. Having produced useful components or assemblies these have to be distributed to users either directly or indirectly. Therefore processes of manufacture culminate in the distribution process with its varied aspects of advertising, transport etc. Finally there is the effect of use, with its feedback to the processes of manufacture.

In the remaining parts of this section, each of these facets will be outlined briefly in respect of its relationship with the main theme.

1.8 RESOURCES FOR PROCESSES

Processes of manufacture are means for using sources of natural energy to convert unmodified natural materials into more suitable forms. Thus every part of such a process, however remote from the world of nature it may seem to be, is a means for using energy that exists in nature. Furthermore the means itself uses adaptations of natural materials and powers. Therefore all the resources of processes of manufacture are drawn from the natural reserves of one sort or another that are all about us.

1.9 APPLICATIONS OF PROCESSES

The study of the applications of manufacture used to be an incidental to a manufacturing technology, such as mechanical or chemical engineering but today, it can be seen as a field in its own right in such disciplines as production engineering and in the work of numerous industrial research establishments devoted to the study of manufacturing methods.

Study of application tends to concentrate in three directions (1) possible uses of a new system of manufacture on the materials in common use (2) feasibility of the use of a process within a particular engineering field (3) design of equipment for applying a suitable process in a particular situation or to a particular material.

1.10 PRODUCTION

The next main branch of processes of manufacture is the use of methods in actual production. This entails at least two and possibly three sub branches of production (1) quality control of the product (2) effective use of the available manipulative processes and (3) possible development of special equipment to adapt the selected processes to the particular selected sequences of manufacture.

Relative cost is a major factor in all production, therefore these three aspects have to be considered essentially from this point of view. Thus the acceptable quality of a finished product is unlikely to be absolute perfection in every detail, but will be a compromise based on how much can be afforded for better quality relative to the absolute minimum of quality that will just enable the product to fulfil its function for a reasonable length of time. The required minimum standard of quality for a reasonable service life must be determined before its manufacture begins and then the product must be checked at a sufficient number of appropriate stages of manufacture to ensure that the standard is being achieved.

1.11 DISTRIBUTION

Finished products or materials are of little use if they are unknown or unobtainable in places where they might be used. Therefore a very important part of processes of manufacture is that related to distribution in its various facets.

In this context advertising, sales service, transport and feedback from users play a significant part. The importance of each of the first three is readily appreciated but the importance of the fourth is not so obvious.

Use can be a process of manufacture in a threefold sense (a) use leads to competition between sources of supply (b) use leads to the development of new uses and (c) use leads to improved standards of living. In the first case the competition generally leads both to improved quality of product and to improved technology and gives a feedback to the development of new processes and equipment. The sophistication that grows with use creates new uses which require new standards of quality and production and again leads to new processes and equipment. Improved living standards create new and greater demands.

BIBLIOGRAPHY

ALEXANDER, W., and STREET, A., *Metals in the Service of Man*, Penguin, Harmondsworth

GORDON, J. E., *The New Science of Strong Materials*, Chap. 1, Pelican (1968)

2 The Nature of Materials

2.1 ENERGY AND ATOMS

In Section 1.3 mention was made of *groupings* of *energy*. This may be a new concept, but it is fundamental to an understanding of materials. Therefore what is energy?

Unfortunately we do not know the basic nature of energy, even if it has one understandable basic nature. All that can be said is that it manifests itself in certain characteristic ways that we can identify and measure. It is known that one type of energy is interchangeable with another in suitable circumstances, but that the total amount of energy remains unchanged and none is ever lost. Some of the recognisable types of energy are (1) electrodynamic (2) electromagnetic (3) electrostatic (4) gravitational (5) inertial (6) magnetic (7) thermal (8) vibrational. Each of these types can interact and can develop physical or chemical forces in one way or another.

What we call a *material* is a zone of influence within which a recognisable orderly association of energy types exists. The pattern enables us not only to distinguish the association by sight and touch but also to utilise it for our own purposes by adjusting and shaping the zone. This adjusting and shaping is done with the aid of other types of energy.

The use of materials is possible only because of the inflexible rules that govern the interaction of specific quantities of the energy types. These interactions conform to certain recognisable patterns of association which tend to repeat themselves as suitable increments of the appropriate types of energy are made available. The basic patterns of energy associations that make up materials are known as *atoms*. Some atoms are stable and can exist for an apparently infinite period of time without their energy state being changed. Other atoms are relatively unstable, perhaps existing for only a few seconds of time before the form of their energy balance changes by the progressive release of some of the original energy in the form of particles or radiation, leaving behind a simpler more stable form of atom. Alternatively, the released energies may themselves form another simpler, stable association to give a third material, also with the release of some energy as radiation or heat. The rate of change or decay of such atoms becomes progressively slower with time in an exponential manner characteristic of the material. It is measured by the *half-life* of the substance, which is the unchanging amount of time taken for the radiation to decrease by half intensity from any given instant.

A well known example of slow decay is that of radium (Ra^{226}) with a half-life of 1622 years, which changes progressively to Lead with the release of a large amount of gamma radiation (a particular form of electromagnetic energy) over thousands of years. Faster decay is shown by polonium Po^{240}) which has a half-life of $0{\cdot}3 \times 10^{-6}$ seconds.

This type of radioactive decay is similar to that shown by *radioactive isotopes*, which are normally stable materials made radioactive by artificially generated internal energy unbalance and used as sources of radiation for industrial purposes. Natural radioactive elements, because of their rarity and cost, are not suitable for such applications. About 100 types of atoms are known, see Table 1 at the end of the book.

2.2 DESCRIBING ATOMIC CHARACTERISTICS

The makeup of atoms is not so simple as might seem from the statements made in the preceding section. For one thing there are always several types of energy associations even in the structure of a simple atom. Some of these types are well known and well understood, although it is doubtful if even one is perfectly understood. Others are little more than identified as being present and some are only suspected of being present and have not been identified. Names are given to the different types of energy manifestation to distinguish them in discussion, but the fact that they are spoken of in a tangible way does not mean that they are tangible to the normal human senses. The limits of the mind make it necessary to talk of these things in terms of an analogy. Thus a *particle* in the atomic sense, describes an effect that is analogous to that of a piece of *real* material but the description may in fact apply only to one limited effect of the particular type of energy association under discussion and it may be equally necessary, simultaneously, to describe the association as a *vibration* or *wave*. The analogue in each case refers to a part of the mathematical law or equation of behaviour that the energy type obeys.

Although this atomic jargon sounds confusing, it is essential to use it to give a comprehensible mental picture of the situation so that ideas can be exchanged and developed.

When it comes to visual illustration the problem is accentuated by conventional symbolism. For example, an atom is often represented as a plain solid ball or, sometimes as a ball with tiny satellites orbiting round it. These are only conventions used to symbolise particular phenomena that form a small proportion of the whole range of atomic phenomena. The intention is not to suggest that an atom is a solid ball, a ball just happens to be a convenient means for symbolising an influence that seems to be located at a specific centre with a zone of effect extending more or less uniformly in all directions from that centre.

It is necessary to accept this use of symbolism, but it is equally necessary to keep in mind its unreality so that we are not deceived into building up a completely false picture derived from our own imagination.

2.3 THE MAIN FEATURES OF AN ATOM

Each of the principal phenomena in the behaviour of an atom relates to one or more of three almost symbolic aspects of the geometry and interaction of atomic structure.

The first aspect is that concerned with those relatively unchanging internal interacting forces which give rise to relatively stable properties such as mass. The effects are considered to operate from a specific concentrated volume called the *nucleus*, set concentric with the middle of the atom.

The second aspect is that concerned with certain less restricted forces, which seem to operate within a more or less symmetrical volume of space surrounding the nucleus. These forces are more directly related to the atom's various reactions with closely adjacent atoms described as physical properties, and chemical properties.

The third aspect is a combination of the other two and is concerned particularly with the more stable interacting forces that may operate between atoms closely grouped together or between groups of one kind of atoms or groups of differing kinds united to give a potentially permanent or relatively lasting agglomeration.

Of these three aspects the third is most important but is not separable from the others when considering the properties of an atomic agglomerate. For simplicity, the principal features of the first two aspects are separated out in this chapter and the cumulative effects are considered in Chapters 3—6.

2.4 THE NUCLEUS OF AN ATOM

The nucleus of an atom is that part that makes up most of its mass and all of the positive electrical potential. The nucleus is pictured as made up of nuclear particles closely concentrated together which each directly contribute to the total properties. These particles are known as *nucleons*, and two types, *protons* and *neutrons* are needed in conjunction with certain short-range forces to explain nuclear properties. In diagrams, the practice of showing nucleons as round balls contained within the larger round ball of the nucleus itself is followed (Figure 2.1) but it must be kept in mind that a nucleus, or

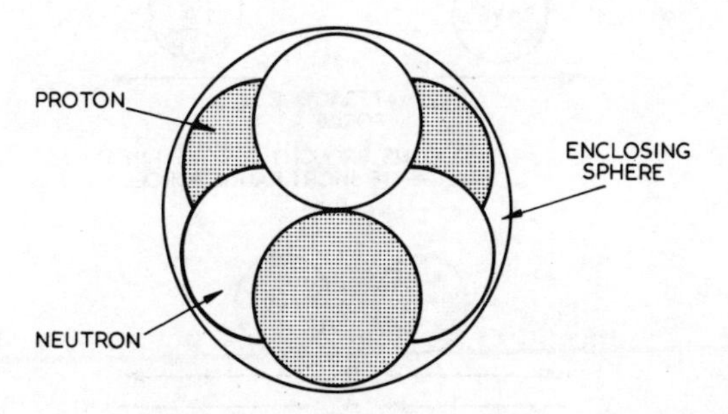

Figure 2.1. When representing a nucleus (e.g. Lithium with 3 protons and 4 neutrons—1 hidden) only the enclosing sphere is shown

even a nucleon is not a simple solid and the ball is meant only as a representation of its presence or of some particular attribute mentioned in the text. The volume of the whole nucleus takes up only a very small part of the total apparent volume of an atom.

2.5 THE PROTON

A *proton* is a nuecleon with certain specific properties, in particular an atomic mass of 1836·1 times the mass of an electron, which has a mass of 9.11×10^{-28} grammes and a positive electric enexgy of 1 *electron volt* (1602×10^{-19} coulomb). Protons make up a large part of the mass and all of the positive charge of any given atom.

Mass itself represents the presence of a very large amount of energy since, energy (E) = mass (m) × c^2 where c is the velocity of light.

The number of protons in the nucleus of a specific atom (called the *atomic number* Z) is constant and determines the number of electrons that are present in a stable atom and hence many of the characteristic properties of the particular atom. It is extremely difficult to separate a proton from the nucleus of an atom and so transmute that atom into another type. This difficulty seems strange since protons have like electric charges and therefore should always tend to repel each other. In fact they do tend to repel each other

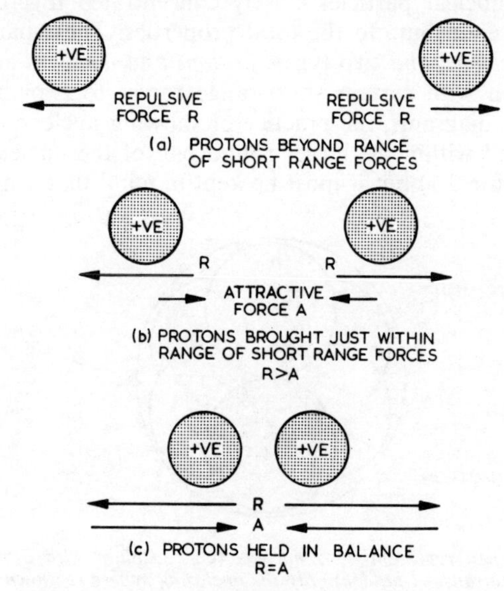

Figure 2.2. Forces on protons relative to distance apart. Note slow increase of repulsion (*arrow length*) *with decreasing distance and rapid increase of attraction. A further decrease in distance past* (c) *would give a large increase in repulsion*

(Figure 2.2a), but when brought within very close range of each other under the influence of externally applied forces, certain short range nucleonic energy effects (Section 2.7) can begin to operate. These effects overcome the natural repulsion and create very strong binding forces which then hold the protons firmly together (Figures 2.2b and c). These forces do not affect the electric potential of the protons. The largest number of protons normally found together in a nucleus is in the atom of the element called mendelevium which has 101 (Z = 101), and the smallest number is 1, in the atom of hydrogen (Z = 1).

Protons are often present in a nucleus in association with another form of nucleon called a *neutron* which possesses almost equal mass. In such cases the relative contribution of the protons is an appropriate part of the total significant mass, an effect which is discussed in Section 2.8.

2.6 THE NEUTRON

A *neutron* has a significant mass of 1838.6 times that of an electron, almost equal to that of a proton but it has no electric potential. As a result, its presence significantly affects the mass of a nucleus but has no detectable effect on its electric charge. The presence or absence of a neutron or even a proportion of the neutrons normally present in a particular stable atom does not appear to have any effect either on the stable number of electrons present in the atom, or on the chemical properties.

Lack of an electric potential means that a neutron can be relatively mobile within the sphere of influence of an atom (it being neither attracted nor repelled by electric charges of either sign). It is held in the nucleus by short-range nuclear forces but it is comparatively easy to add or remove a neutron from a nucleus. A neutron cannot be

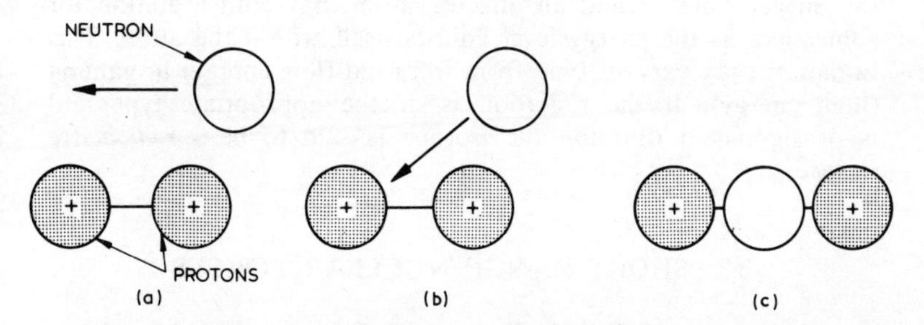

Figure 2.3. A neutron is unaffected by electropotential but can be held by short range forces. (a) A neutron passing unaffected near to a pair of protons coupled by short range forces. (b) A neutron impacting at low energy on a similar pair of protons. (c) The neutron captured by short range forces

bound into any part of an atom other than the nucleus, therefore if a neutron is impelled into an agglomeration of atoms it will tend to pass straight through without loss of energy simply because the space between nuclei is great. If it happens to strike a nucleus it may be deflected, perhaps knocking out another neutron already present in the nucleus, or it may itself be absorbed. (This is one reason why bombardment by neutrons is a useful tool in the study of nuclear physics.) See Figure 2.3.

The number of neutrons (N) that can be present in a given nucleus does not seem to be precisely fixed but does seem to be closely associated with offsetting the effects of the increasing mutual repulsion of protons as the number of the latter rises within atoms of increasing atomic number. The maximum appears to be about 1·5 times the number of protons in atoms with large atomic numbers and about equal with the protons for atoms with smaller atomic numbers. The minimum for a given large atom is not known, but variation of the number over a characteristic narrow range does not seem greatly to affect the properties of the relevant atom other than to change the energy balance somewhat and to influence the mass. There is a statistical average number of neutrons typical of each given type of atom. Any atom containing a number of neutrons outside the variation range of the average is known as an isotope of the typical atom. The atoms of hydrogen are the only ones that are normally free of neutrons. Isotopes exist naturally in most materials but may also be created artificially by bombardment with neutrons. This bombardment can either cause ejection of neutrons from some nuclei, without replacing them, or can add to the number in others depending on the energy of the bombardment. Alteration of the number of neutrons from the average in an atom does change the energy balance and an affected atom may emit radiation for some time as the energy level adjusts itself within the atom. This radiation may vary in type from infra-red (low energy) to gamma (high energy). If the radiation is of the appropriate type and is of significant duration the isotope is said to be a *radioactive isotope*.

2.7 SHORT RANGE NUCLEAR FORCES

It is obvious that if similarly charged protons are to be held together in a nucleus then their natural electrical repulsion for each other must be overcome, see Figure 2.2. The type of forces that produce the necessary effect are called short-range nuclear forces and it is not known if they are an independent influence or simply a little-understood aspect of proton behaviour. Some workers identify these forces as nuclear particles known as *mesons* but too little is known about them to be completely certain about their nature.

All that need be stated here is that there are certain forces that perform two functions: (1) holding protons in certain strong, lasting relationships with each other without affecting their positive electric potentials or their masses, and (2) holding an approximate number of neutrons in a close association with a characteristic grouping of protons.

2.8 THE MASS OF AN ATOM

Most of the total mass of an atom is made up by the sum of the masses of the protons (Z) and neutrons (N) in a nucleus. Only a very small part comes from other atomic particles and effects.

The *atomic mass number* or *atomic weight number* (A) for a particular atom is the ratio of the mass of that atom compared to the mass of the isotope of carbon (C) which contains 6 protons, 6 neutrons and 6 electrons regarded as 12 units, thus giving the *atomic mass unit* (AMU) a weight of $1{\cdot}65979 \times 10^{-24}$ grammes. A is usually quoted as a superscript to the atomic symbol, e.g. C^{12} for this carbon isotope. The value of A for an atom should come out nearly equal to a simple integral of the mass of a proton or the mass of a nucleon since these masses are nearly equal to each other and the mass of the atom's electrons is very small. Hence if the value of A is known for an atom the difference between that number and the atomic number Z gives the likely number of neutrons N. For example, for the element beryllium (Be^9) $A = 9$ and $Z = 4$, so N for beryllium is 5. Similarly iron ($Fe^{55{\cdot}85}$) with a value of $A = 55{\cdot}85$—say 56—and the value of $Z = 26$ has a value of $N = 56 - 26 = 30$ neutrons. The value of Z is often quoted as a subscript to the element symbol so beryllium could be given as ${}_4Be^9$ and iron as ${}_{26}Fe^{55{\cdot}85}$.

Kinetic energy if an atom is moving at very high velocity can contribute a very small addition to the mass but the effect is completely negligible in the present context.

The contribution of electrons, see next Section, can be more significant but is still quite small and is not very important to materials technology.

2.9 ATOMIC SHELLS AND ATOMIC DIAMETER

The second aspect of interest in the study of the atom is the almost empty space surrounding the nucleus, within which the energies of the nucleus react with the remaining most important atomic particle, the *electron* to produce the equivalent of a nearly constant spherical volume of influence sometimes described as an *atomic shell* or *electron shell*. Although this shell is almost empty, the electrons contained within it move about so rapidly that to an observer able to use a microscope capable of magnifying 10^9 times, it would look full. In fact an agglomeration of atoms looks solid for this same reason—the rapid movement of the electrons. In addition the electromagnetic forces of the moving electrons react with similar forces in other atoms brought near to them so an agglomerate feels solid as its atoms react with the atoms in the finger tips. The average diametral distance over which the shell influence of a specific atom works is called the *atomic diameter*.

That solid matter is largely empty space is shown by the way in which light rays pass through transparent materials and gamma or X-rays pass through greater or lesser thicknesses of opaque materials. In fact the ease with which the differing types of electromagnetic vibrations pass through a material or are diffracted or stopped gives a clue to the nature of the influences operating in the individual atomic shells since the observed effects are almost entirely those of interacting wave forms. That is, the projected rays are interacting with the electron oscillations within the atomic agglomerate.

2.10 THE ELECTRON

An *electron* is a particle of very much smaller size than a proton or a neutron and 1/1836 of the mass of the proton. It possesses a large electric potential numerically equal to that of the proton but of opposite sign and has a high level of kinetic energy.

Because of the respective electric charges the number of electrons must equal the number of protons in any given atom if that atom is to be electrically neutral. Thus a stable atom always contains a constant equal number of protons and electrons (the atomic number Z). However, owing to the dynamic characteristics of electrons and their relative freedom of movement within a parent atom this neutrality is the statistical average value of a pulsating charge rather than a simple state of neutrality.

If a free electron is unaffected by outside influences it will move in a straight line in space. Application of a magnetic or an electric field to a moving electron causes it to change its path in accordance with the normal laws of electromagnetic force. Thus, if it is captured by an unbalanced positive field of a proton or a nucleus, an electron will be drawn into a form of orbital motion around the centre of attraction. The actual path of travel around a nucleus although controlled by the basic laws of mechanical oscillation (wave mechanics) will be so complex and rapidly variable that it is impossible to visualise it in terms of human experience. For convenience of visualisation an electron in an atom is often shown as a satellite performing a simple orbit around the nucleus, as in Figure 2.4a, but this picture is much too simple for reality. The movement is better represented as a cloud around the nucleus (Figure 2.4c).

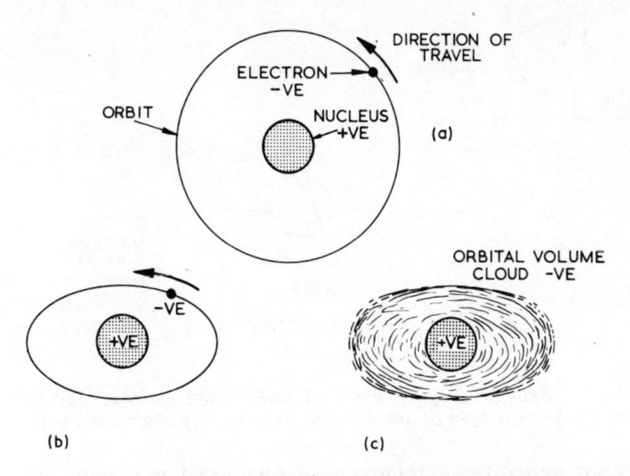

Figure 2.4. Electron behaviour; (a) circular orbit; (b) elliptical orbit; (c) ellipsoidal orbital or zone of probability

This cloud, called an *orbital*, represents the zone within which the relevant electrons might be found at a given instant. It is simply a zone of *probability of presence*. Change in the density of shading of an orbital may be used to indicate greater or lesser probability of the electron being present at a given instant. It will be apparent from this concept of the orbital that the effect of an electron's negative potential will be greater where probability of its presence is greater and less where the probability of its presence is less, so the actual effect will be one of a locally pulsating charge. Hence a zone of low probability will tend to be electropositive and a zone of high probability will tend to be electronegative. In a stable atom these effects will average each other out in relation to the total effect on the relatively constant positive field of the nucleus.

Because of the critical importance of the orbital aspects of electron behaviour, orbitals are given more detailed consideration in the next section.

The picture of individual electron behaviour is not yet sufficiently complete to go on to other effects because there is another phenomenon which greatly influences the behaviour of an electron in an atom. This affect is the *spin* of an electron. A captured electron not only travels about within an orbital zone in an atom but simultaneously appears to spin about its own axis as shown in Figure 2.5.

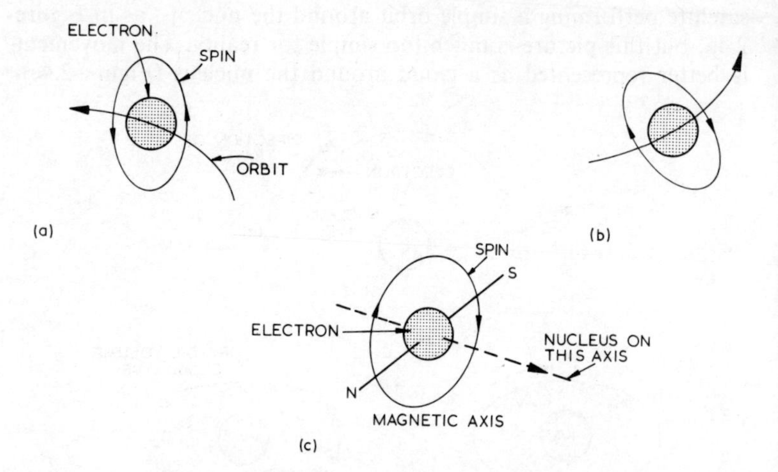

Figure 2.5. Spin of a single unpaired undisturbed electron. (a) Spin at one part of orbit; (b) Spin at another part of orbit; (c) Magnetic effect of spin

This spin generates a relatively strong local magnetic field which greatly influences the electron's behaviour within its orbital and may also affect the magnetic properties of strongly magnetic materials.

2.11 ELECTRON ORBITALS

The behaviour of an electron when it moves into the range of influence of a nuclear orbital is the result of the interaction of many factors. The most important are (1) the presence of other electrons already in the system (2) the effect of thermal excitation (3) the interacting forces from adjacent atoms that may be present.

As suggested in the preceding section, electrons in a state of oscillation within the orbitals associated with an atom have to obey the rules of wave mechanics. Because stable vibration modes cannot persist outside the range of certain discrete resonant energy levels, or *quanta*, the possible number of basic energies of vibration in an atom depends on the total attractive energy of the protons in the nucleus. Thus, orbiting is possible only in certain modes within certain energy zones known as *Principal Quantum Numbers* or sometimes as *Principal Shells*, beginning at PQN 1, with lowest energy, increasing in definite energy steps, in strict relationship to the relevant energy of the nucleus, until the rarely-occupied highest possible normal energy level is reached at PQN 7 for large atoms. Alternatively these shells are sometimes given identifying reference capital letters running from K to Q. Within each energy level only a specific number of modes of stable electron movement is possible (as in any resonating system) since otherwise, the electrons would interact violently with each other.

A maximum of only four modes of electron movement are possible within a Principal Quantum. The lowest Principal Quantum Number has only one of these modes, PQN 2 has two, PQN 3 has three and PQNs 4 and 5 have four modes each. Each of these modes, sometimes called a *Secondary Quantum Number* or *subshell*, has the characteristic shape of basic electron orbital pattern, or cloud of probability, within which not more than two electrons may oscillate at one time. A specific maximum possible number of orientations of orbital pattern, each pattern being completely occupied if two electrons are in it, is possible for each mode, the number being determined by the relevant Principal Quantum Number. It should be noted that the laws of wave mechanics make it impossible for an electron to settle into any one particular state within an orbital until all lower energy states aremple cotely filled. At higher energy levels some modes within a lower energy band may require higher energy than some within the next higher energy band. The four types of orbital mode are identified in order of increasing energy, by the small letters s, p, d, and f, s being the simplest mode and f the most complex. Thus a particular orbital may be described as (3s) which is an orbital of an energy level governed by the Principal Quantum Number 3 and of the type s. The possible maximum number of electrons conforming to the pattern may be added as a superscript, $(3s)^2$ indicating that in that particular orbital not more than 2 electrons can ever be found (never more than 2 are found in any s state, see Figure 2.7). The maximum level to be occupied is determined by the total energy of the nucleus and it is the characteristic pattern of orbital shapes.

The maximum normal energy level (outer shell) is the one whose electrons exert an almost exclusive influence on the physical and chemical properties of the particular atom. Therefore, because the similarity of orbital characteristic imposed on differing atoms by the laws of wave mechanics may give them similar outer shell characteristics, one might expect similar properties from certain otherwise very different atoms. This similarity is in fact found to exist, hence the significance of the *periodic table* for classifying atoms.

If the level of energy possessed by an electron is near or above the level of that of the limited maximum attractive energy of a nucleus, the chance of that nucleus catching and holding such an electron in any of its empty or partly filled outer orbitals is small, even if the number of possible orbitals is high. For this reason, some possible orbitals of overlapping energy level in the maximum principal shells particularly at the PQN6 and 7 levels are likely to remain unoccupied even in normal stable conditions. In any case an incomplete maximum Principal Quantum Number can never contain more than eight stable electrons until the next higher Principal Quantum Number has begun to fill. Thus atoms with high maximum Principal Quantum Numbers tend to be more unstable than atoms with lower maximum numbers, hence the unusual properties and possible short life of atoms with atomic numbers in the 90 plus range.

It is easiest to visualise successively higher energy orbitals as thick walled hollow spheres successively enclosing each other as in Figure 2.6. The radius then represents the magnitude of the energy

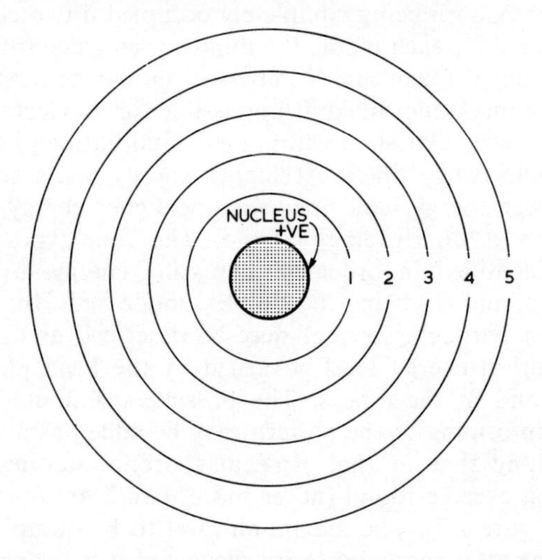

Figure 2.6. Representation of energy levels of Principal Quantum Numbers as shells

level and the thickness the spread of energy. This picture is reasonably true for the first one or two shells but becomes increasingly inaccurate at higher levels as the number of modes increases. At levels 3 and 4 overlapping of orbital energy begins; orbital level 4s being of lower energy than 3d. Hence, although at the lower quantum levels no electron can become stable in a given orbital until all the available lower energy orbital levels are filled, at nominally higher quantum levels, owing to such orbital energy overlap, less regular behaviour occurs, giving rise to *transitional elements*. Thus the element Iron (Fe—atomic number 26, Table 2) has 2 electrons in the 4s level but only 6 of a possible maximum of 10 in the 3d level. In Figure 2.7 is indicated the maximum number of electrons

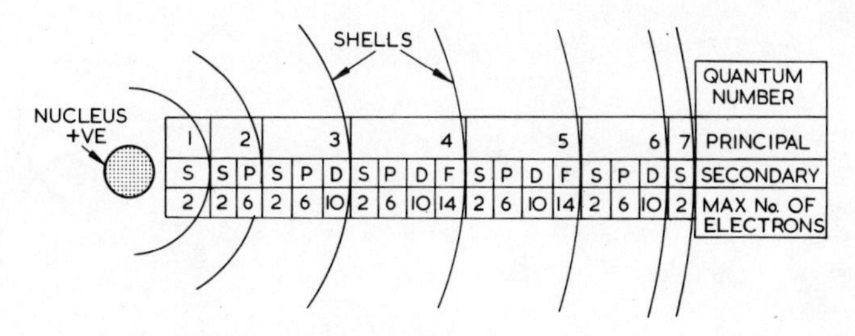

1	2		3			4				5				6			7	PRINCIPAL
S	S	P	S	P	D	S	P	D	F	S	P	D	F	S	P	D	S	SECONDARY
2	2	6	2	6	10	2	6	10	14	2	6	10	14	2	6	10	2	MAX No. OF ELECTRONS

Figure 2.7. Maximum possible number of electrons in each Quantum state. Correlate with Figure 2.6 and Table 2

possible in each type of orbital for each Principal Quantum Number. Table 2 at the back of the book gives the electron disposition in each of the elements up to Radium ($Z = 88$).

Electrons tend to pair with each other within particular orbital types because a pair is more stable. Pairing is more stable because the two electrons repel each other tending to keep positions at opposite sides of the nucleus at any given instant, thus reducing the overall energy difference in the atom as shown in Figure 2.7 and their spin axes settle in parallel opposition to each other to form a stable closed magnetic field, see Figure 2.9b. Some magnetism is

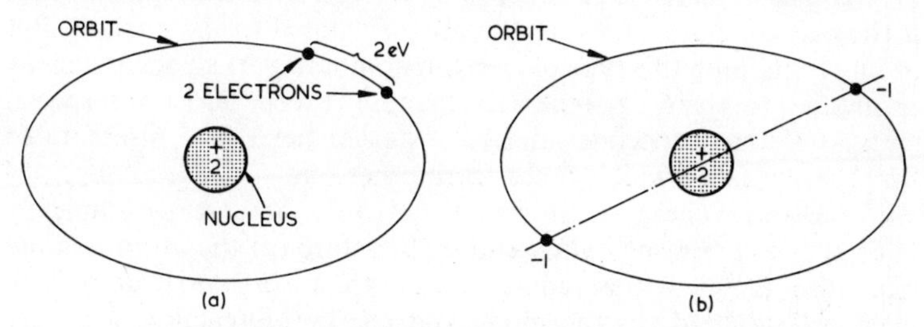

Figure 2.8. Electropotential forces encouraging pairing of electrons in a shared orbit. (a) Two adjacent electrons, maximum energy difference 4 eV (+2 − (−2)). (b) Opposite electrons, maximum energy difference 3 eV (+2 — (—1))

generated by the orbital movement of the electrons but its effect is generally very slight and may be ignored in the present discussion. If for some exceptional reason, two such paired electrons within an atom spin on the same relative axis direction, a strong external magnetic field is set up and that particular material will have strong magnetic properties.

A single unpaired orbital electron sets up a more unbalanced pulsating electrodynamic field (Figure 2.9a) than two paired electrons, therefore adjacent atoms with the former condition prevailing

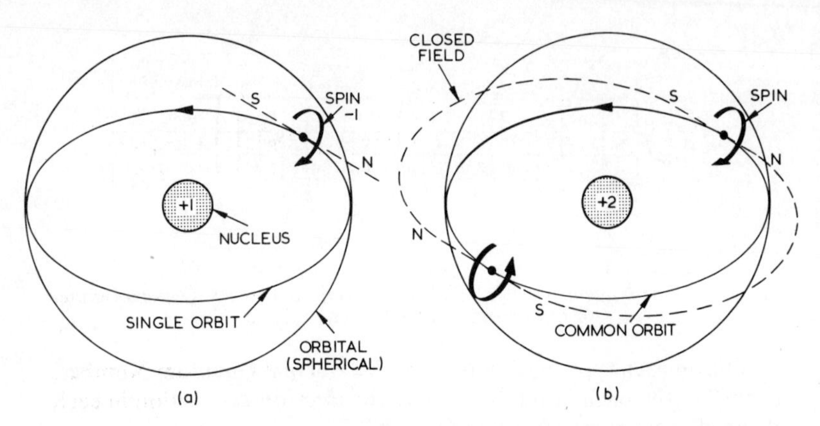

Figure 2.9. Magnetically neutral stability of paired electrons of opposite spin. (a) Single electron in Hydrogen atom giving a random magnetic field when unaffected by outside influences. (b) Closed magnetic field between 2 electrons paired in an atom of Helium

in an outer, or a high-energy next-to-outer shell (this state cannot exist in a low energy inner shell of the same atom), tend to react more strongly with each other than do atoms in other conditions.

A single unpaired electron has a strong random magnetic field, if the axis of spin is not directionally orientated to the nucleus. But atoms containing this type of electron do not impart strong magnetic properties to their agglomerates because, except in rather special situations, the respective adjacent external fields will orientate to form closed fields which cancel each other out.

If the outer-energy-level orbitals of an atom are incompletely filled they are not inherently stable, therefore (a) the atom is more likely to interact with an adjacent atom than is a similar atom with filled orbitals and (b) the mode and energy of reaction will vary with the degree of imbalance of the initial atom related to the imbalance of the adjacent atom or atoms. This is why the chemical and physical properties of an agglomerate are so closely associated with the outer orbital electrons—the so-called *valency electrons*.

Study of the actual shape of an orbital within an atom is a complex subject, but it may be mentioned that in basic isolated stable conditions, an *s* orbital is spherical in shape whether occupied by one or a pair of electrons, see Figure 2.9. A *p* orbital is hour-glass shaped for each pair of electrons, see Figure 2.10, and a p sub-shell is completed by three orbiting pairs of electrons with orbital axes mutually at right angles. An atom with an outer shell filled to both *s* and *p* level would have an overall orbital shape rather like a sphere with six knobs projecting from it as shown in Figure 2.10c. Of course,

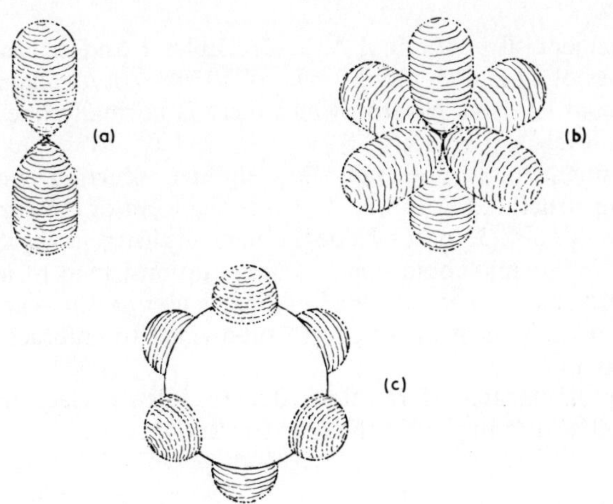

Figure 2.10. Types of orbitals (a) p type orbital axis followed by 1 or 2 electrons (b) A set of 3p axes with 5 or 6 electrons (c) filled s and p orbitals (s is spherical)

the probability of the presence of an electron would not be exactly the same throughout the whole volume of the total orbital, but such variation cannot be shown readily in a two-dimensional representation. The d and f orbitals are far too complex for discussion in the present context. The actual type of outer orbital in an atom greatly affects its mode of reaction with other atoms and helps to explain the definite ordering arrangement of atoms that is found in some groupings, notably crystals (see Chapter 3). It should be kept in mind that the orbital shapes considered above apply only to an independent stable atom unaffected by adjacent atoms. If other atoms are present then the shapes are likely to change to accommodate the interacting forces.

2.12 THE PERIODIC TABLE

Once a stable atom has formed with its full complement of protons, neutrons, electrons and other force agencies it has overall properties and physical characteristics that are unique. Another type of atom may possess similar properties but never a set completely identical in all respects. The balance of forces in the stable atom is very difficult to upset and cannot be changed by anything other than extremely large external forces. Each type of atom is called an *element* and is given a unique identifying name and description which includes its atomic number, its atomic mass and the disposition of its constituent electrons in relation to its Principal Quantum Numbers. For example, the element silver (symbol Ag), see Tables 1 and 2, has an atomic number of 47 and an atomic mass of 107·88. All its quanta are filled to the 4d level (46 electrons) and there is normally one electron in the 5s level.

A simple table does not show up the recurring similarities of atomic structure but can be drawn in the form of the *Periodic Table* shown in Table 3. In this Table the elements listed in vertical columns have similar maximum quanta configurations, tend to have similar properties and so are included in a particular *period*. There are eight main periods most of which are subdivided to embrace more than one column.

The similarities of reactions due to valency electrons are considered a little further in the next subsection.

2.13 VALENCY ELECTRONS

As already suggested, a stable atom with completely filled Principal Quanta is likely to be so stable in behaviour that, except under unusual circumstances it will be unaffected by the energy fields from other, even closely placed, atoms. Thus, reaction between adjacent atoms is only likely to occur if the atoms concerned have each an incompletely filled highest Principal Quantum. Incompletely filled second highest Principal Quanta may also contribute something to a reaction, notably in respect of magnetic effects. Reactions between atoms are closely associated with the relative degree of pulsation of electropotential balance within the individual atoms (see Chapter 3).

It follows that a knowledge of the behaviour of electrons in the maximum Principal Quanta is of particular importance to the understanding of reaction between atoms and, in particular to the study of the bonding of atoms with one another as they form coherent substances. Since outer orbital electrons have a marked connection with the strength of bond formed between adjacent atoms, they are called *valency electrons* (from the Latin word for strength) and an atom's *valency* is its power to link firmly with other suitable atoms as indicated by the number of electrons in its highest energy Principal Quantum.

The maximum number of electrons continuously in the highest Principal Quantum of any one atom can never be higher than 8, see Table 2. A valency number is not a simple quantitative representation of the strength of bond one atom forms with another; however, the number can indicate the potentiality that an atom has for linking up strongly with particular maximum numbers of other atoms of the same or different types according to the nature of the bond, or bonds, that can form (see next chapter). A valency of 1 suggests that the atom possessing it is not likely to join strongly with more than one other atom at a time, a valency of 2 may suggest the possibility of linking with two other atoms, but this could be upset by other factors. Thus a hydrogen atom, valency 1, is never found linked to more than one other atom although more than one hydrogen atom may link separately to the second atom, if the latter is suitable, as in the water molecule (H_2O—two hydrogen atoms linked separately to one common oxygen atom in each molecule). Helium (He) possesses a valency of 2 but these two electrons completely fill the Principal Quantum Number 1 which is thus inherently stable; consequently, a helium atom is never found firmly bonded to another atom. Magnesium (Mg) also has a valency of 2, but this time from an unfilled Principal Quantum Number 3, so magnesium can bond readily with two other similar atoms. With 8 valency electrons (elements Ne, A, Kr, Xe, Rn) the maximum Principal Quantum is either completely filled or is relatively stable and the likelihood of strong bonding is small. A valency number over 8 is possible in some transition elements when two outer Principal Quanta overlap.

A chemist is interested in complex molecular structures and to study them he sometimes uses a diagrammatic system in which each valency link is represented by a small bar, as in Figure 2.11, which

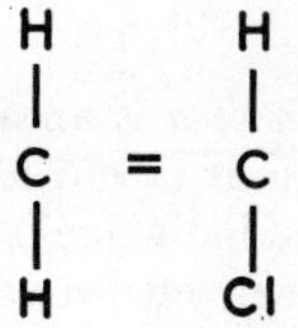

```
H     H
|     |
C  =  C
|     |
H     Cl
```

Figure 2.11. Representation of valency links in a molecule. Note the double link formed between the carbon atoms in this example of the vinyl chloride monomer

shows the basic system of the vinyl chloride molecule. Note the double bond between the carbon atoms (C) giving each carbon atom a total of four links (carbon valency is 4).

The materials technologist is interested in the relative strength of valency bonds and in their effects on properties. Hence the next Section and Chapter 3 are concerned with the states of matter and with the nature, strength and characteristic features of the bonds commonly found in structural materials.

2.14 THE STATES OF MATTER

Matter is found in three familiar states: gas, liquid and solid. Although these states are usually distinguishable there are many variations, particularly in the liquid and solid states, so it may sometimes be difficult to say which state a material is actually in.

The normal distinctions between states of matter and between variations within them, are dependent on the nature of the interatomic reactions that are operating between the constituent atoms. These reactions are rarely caused by the operation of one solitary mechanism, but are the result of a complexity of mechanisms. An understanding of these mechanisms gives one of the most important keys to the understanding of materials and they are discussed in Chapter 3.

It is desirable first of all to define what is meant by the terms gaseous, liquid and solid.

The gaseous state is one in which the constituent atoms, or basic molecules have no close rigid association with each other and are in a state of continuous random motion, appearing to repel each other in such a way that they spread out within the limits of any confining vessel, thus generating a pressure. Ambient conditions greatly influence the behaviour of a gas and a *perfect* gas would obey the law $pV = RT$, when p is the pressure in the gas, V is the total volume, R is the appropriate Gas Constant and T is the absolute temperature. However for various reasons associated with atomic structure, no gas is perfect, so the law has to be modified (to fit with the findings of J. D. Van der Waals) to give:

$$\left(p + \frac{a}{V^2}\right)(V - b) = RT$$

where a is an allowance for certain weak attractive forces between atoms or molecules and b is an allowance for the actual volume of the individual atoms or molecules present in the total volume. These weak attractive forces between atoms, known as *Van der Waals forces* or *Van der Waals bonds* are attributable to electro-potential attraction between atoms resulting from the pulsating nature of their respective electropotentials.

The *liquid* state is that in which the constituent atoms or molecules of the material are held lightly, but not rigidly, in contact with each other in a state of continuous sliding movement or *diffusion* past each other at a speed much lower than their speed of random movement in the gaseous state. The volume of a liquid is about equal to the total volume of the atoms or molecules in it and, under the force of gravity will change its shape to fit a confining vessel, see Figure 2.12b. A liquid responds to ambient conditions and obeys the same modified gas law but with the constants appreciably changed. The factor a/V^2 is much larger and the factor $(V - b)$ is very small. It can be seen from these differences that pressure has much less effect on a liquid than it has on a gas.

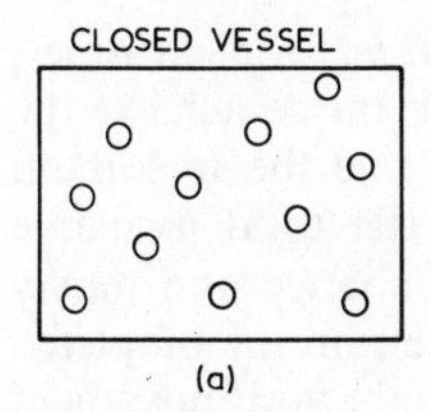

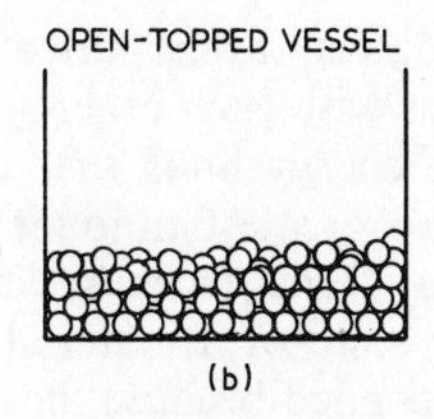

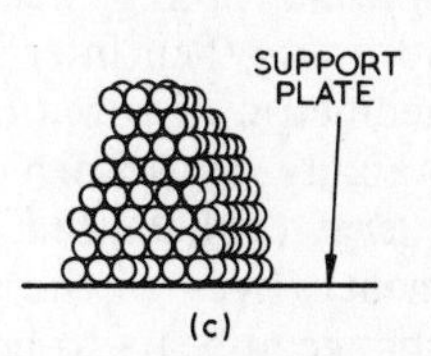

Figure 2.12. The basic states of matter. (a) In a gas the atoms or molecules (represented as circles) are continually rebounding from each other and any container wall. (b) In a liquid there is a slower sliding movement influenced by gravity. (c) In a solid there is little movement and the total body tends to be self sustaining

The *solid* state is that in which the constituent atoms or molecules form such strong short-range bonds between each other that they tend to stay in constant positions with respect to each other in the form of an agglomerated mass. A solid is self-sustaining, has shape as well as volume (Figure 2.12c), and exhibits mechanical properties that can be sufficiently significant for the material to be used as part of a structure. Ambient conditions affect certain properties of a solid, in a similar manner to the way they affect a liquid but in particular, they may affect the mechanical properties in ways that can be most important to the materials technologist. In the latter respect some effects (of heat treatment strengthening a material, see Chapter 6) are highly desirable and others (incipient melting by over-heating, see Section 2.16) are barriers to effective use.

In differing ambient conditions many materials can exist in all three of these states transforming from one state to the other with the appropriate change in ambient conditions. The occurrence of such a transformation is usually distinguishable, particularly with the aid of suitable instruments, but there can be exceptions to this as, for example, when the liquid state is very viscous and the solid state very weakly bonded. Certain complex materials may be able to exist only in one state and may disintegrate or lose their identity if change is attempted.

It is also possible for a material to be solid or liquid in more than one clearly recognisable way in particular ambient conditions (see Chapter 3). Each distinguishable condition of a material is called a *phase* and this term will generally be used in this sense in the subsequent text. If a material, able to exist in different states, is normally either a liquid or a solid, it is customary to call its gaseous condition, if it has one, its *vapour* or its *vapour phase* to distinguish the condition from that of a material that is normally a gas (carbon dioxide is a gas but steam is water vapour).

Not one of these changes of state can be accomplished without *diffusion* taking place. That is, atoms move relative to each other, adjusting their interactions with each other and trying to achieve the geometric distribution of energy links that will give the maximum stability by absorbing the greatest amount of the total available energy (unbalanced energy tends to cause atom change and movement). Even in the solid state of a material when an uncompleted change appears to be suspended because individual atom movement is so difficult, some diffusion movement still goes on, not only at outer surfaces, see Section 2.16, but also within the material's structure. Movements of this kind take place at rates controlled by (1) the nature of the material, (2) the internal state of the material, and (3) the environment. In some cases, atom movement is so slow and so few atoms are taking part, that the effect can be ignored and the material is said to be in a *metastable* state. In other cases, movement may be so rapid that significant structural changes are caused in a short time.

Each state of a material, and each phase within each state, is characterised by certain physical properties. Such properties group themselves into three divisions, (1) mechanical properties—reactions related to externally applied mechanical forces, (2) thermal properties—reactions to thermal influences and (3) electro-magnetic properties—reaction to applied electrical and magnetic forces. The first division is introduced in the next Section, but all are outlined in Chapter 4 and are considered sectionally where relevant in other parts of the text.

2.15 MECHANICAL PROPERTIES OF MATTER

The different mechanical properties of a material are each an indication of some aspect of cohesion of the atoms and molecules within the material. These properties are usually measured in relation to the application of one or more of three general types of external force.

(1) Tensile force which tends to pull the atoms apart in one direction
(2) Compressive force which tends to force the atoms closer together in one direction
(3) Shearing force which tends to cause the atoms to slide past each other

Not one of these general types can be applied in isolation to a material, because the application of one implies some application of the others, even if only in a minor intensity. So the mechanical properties of a material are a compromise between the ideal individual interactions predicted by atomic theory.

Mechanical properties of a solid material need not depend only on the types of bonds acting in the material but also on the geometric arrangement or local grouping of atoms or molecules in relation to the kind of force being applied. Strength may be imparted not only by bonding between molecules but also by geometric interlocking of suitably shaped molecules on the same lines as the interlocking of the fibres in a rope or the links in a chain. The individual fibre strands of a rope are short and normally unbonded to each other (see Figure 2.13). If they are intertwined and geometrically

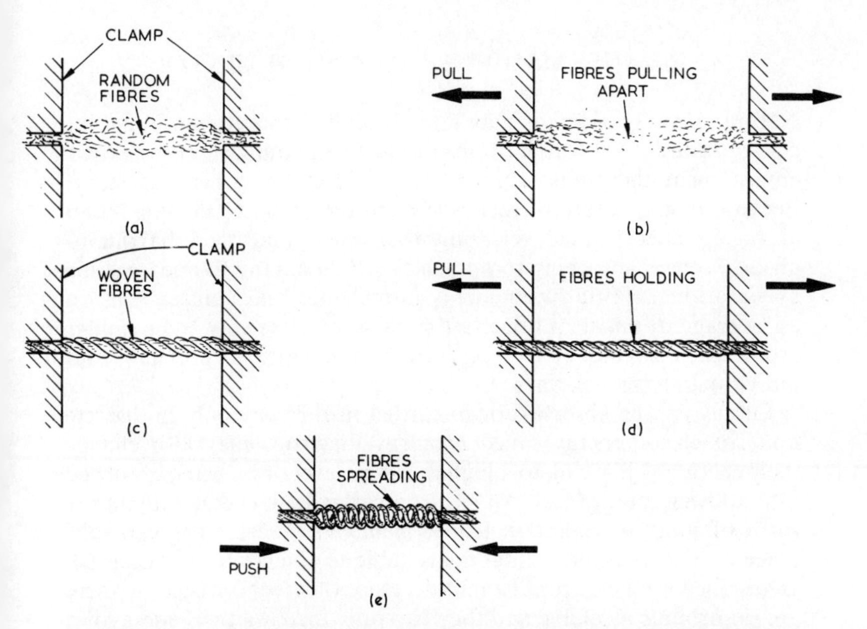

Figure 2.13. Mechanical interlocking of fibres. (a) Random mixture (as in cotton wool). (b) Little strength shown by fibres in (a) subject to tension. (c) Woven or spun fibres (as in rope). (d) Strength shown by rope fibres in tension. (e) Low axial strength of rope fibres in compression

interlocked with each other in a systematic way (Figure 2.12c) the strong bonds within the independent strands are able to exert their strength to resist fracture when a load is applied (Figure 2.12d). If a load is applied to a rope in an unsuitable way, such as axial compression (Figure 2.12d), the inherent weakness of bonding between the strands is shown up. Geometric effects of a similar kind are commonly found within the structure of composite materials of many kinds. Thus it may be possible to give strength to an inherently weak material by rearranging its atoms or molecules in a more effective geometric array or by suitably mixing it with another, perhaps more fibrous material, particularly if some bonding between the materials is also possible. Long curly molecules (see Section 3.16) are often amenable to treatments like these.

Mechanical properties in any particular material depend on the inter-relationship between the following influences acting in its internal structure. (1) Strong short-range interatomic forces. (2) Weak long range interatomic forces. (3) Geometrical interaction between the different particles, phases or molecules.

2.16 THERMAL ENERGY AND STABILITY

Nature always tries to maintain equilibrium between adjacent energy states. Therefore if atoms are to be in equilibrium with their environment they must absorb or release energy when necessary. Because of the nature of their structure they tend to do this by an *averaging* effect; some releasing too much and then having to absorb, others absorbing too much and then having to release some as vibrational elastic strain energy interchange takes place. Thus on an average atoms at n.t.p. must possess more energy than similar atoms at absolute zero temperature and normal pressure, but individual atoms will vary.

Energy can be absorbed or discarded in this way only in discrete units of elastic energy called *phonons*. Electrons may also change their energy by jumping to a higher energy level or by being absorbed into a lower energy level. An electron can release energy only in the form of units of radiation (called *photons*). These steps can take place only if a vacant orbital is available to the electron concerned. Inner filled shell electrons cannot decrease their energy because there are no orbitals available and they can only increase their energy if a higher energy orbital is available. Thus only the upper quanta level, or next-to-upper quanta level electrons are likely to be free to jump up (such electrons in an atom are often called *free electrons*) and only empty or partly-filled outer quanta levels are open to receiving more electrons. The nett effect as energy is absorbed, is that (a) the outer quanta become less precisely and constantly filled, (b) higher-than-normal quanta begin to hold electrons (c) a continual exchange of electrons is likely to go on between the outer quanta (d) some electrons may be ejected altogether leaving vacant orbitals and (e) electrons from other atoms or sources may be captured in vacant outer orbitals.

With absorption of thermal energy atoms appear to become

larger as their charges pulsate more randomly. Increasing the temperature of a solid causes more and more vigorous vibration between atoms until the stable positions normal in the solid state are no longer tenable and the solid begins to *melt*. Further increase in temperature causes still more vigorous vibration and some atoms escape from the surface. More and more atoms escape as the temperature rises until there are more leaving than dropping back and the substance begins to *evaporate*. Eventually, the *boiling point* is reached and the material changes to the *vapour state*. Certain materials may go straight from the solid to the vapour and these are said to *sublime*. Falling temperature causes these effects to reverse and provided that the vapour has not dispersed, it will *condense* back into the bulk liquid state (or directly to the solid state if appropriate) and, with further fall in temperature *freeze* into the solid state.

Thermally induced changes of state are always progressive from certain localised zones or centres called *nuclei* (not to be confused with atomic nuclei). Progression is in a direction appropriate to the prevailing conditions of energy interchange.

Since at n.t.p. there is always some thermal energy causing atoms to vibrate there is always some outward movement of the surface atoms of a solid and some surface evaporation from a liquid.

Change in the thermal reactions between atoms within a solid material may cause them to reshuffle themselves into other stable geometric arrangements at different temperature levels, giving different states of solidity (differing solid phases). Such changes usually involve a dimensional change in the material, sometimes a contraction in opposition to normal thermal expansion with rising temperature or vice versa. If these changes can occur without the chemical identity of the material being lost the material is said to show *allotropy*. Most allotropic changes are reversable at about the same temperature level, but others may persist after temperature has reverted until sufficient energy is accumulated to cause the change. In materials containing more than one type of atom, similar but irreversible atomic rearrangements may occur accompanied by an irreversible change in chemical identity. In these cases a new material with its own peculiar properties is formed.

Because thermal excitation of an atom has its most obvious effects on the valency electrons, chemical reaction between different atoms usually becomes easier at higher temperature which explains the speeding up of many chemical reactions in materials when heat is applied.

If thermal energy from external heating is not completely absorbed within a structure what happens to the surplus energy? Valency electrons are continually being transiently excited into higher than normal energy states and then, as they fall back into their normal stable states the surplus energy is released in the form of photons in discrete wavelengths corresponding to the magnitude of the energy change. Thus low energy release is detected as invisible low frequency infra-red radiation. With increasing energy release the radiation begins to include frequencies that rise into the visible spectrum showing first red, orange, yellow, then white radiation as from so-called whitehot metal. Heating beyond this intensity can lead to outer electrons being transiently ejected from the mass of atoms in *thermionic emission* a phenomenon much used in electronics.

Thermal energy is transferred through a solid material primarily either by transfer of valency electrons, when that is possible, or by progressive resonating between the valency electrons of adjacent atoms. It is impossible for all the atoms inside a material to react simultaneously to a change in external thermal conditions. Time must be allowed for the spread of the change. Thus specific basic changes of behaviour that occur at fixed thermal energy levels on the individual atom scale become blurred into graduated changes in a mass of real material. In a liquid, heat transfer is assisted, in part, by the random movements of the atoms about each other developing into more specific *convection* circulation movements under the influence of density differences caused by temperature differences.

2.17 PRESSURE EFFECTS

Pressure has effects that are complementary to those of temperature. Pressure arises from atoms reacting against each other when they are restrained into a space they would not inherently occupy. Remove atoms from any such space and the pressure drops, add atoms and pressure rises. Absolute zero pressure could only occur in the complete absence of all atoms.

The presence of free atoms near to each other influences the thermal balance in any space. Thus if atoms of a gas are forced closer together, from a previously thermally stable state, by applying increased pressure, their thermal energies are no longer in equilibrium (orbiting electrons have too much kinetic energy for the new conditions). Therefore the gas heats up to a temperature transiently above the temperature of the containing vessel, until the surplus thermal energy has conducted away through the walls. If pressure is reduced the reverse effect takes place. A liquid responds to pressure in a similar manner, but the basic interatomic forces are very much greater so the influence of pressure on thermal energy is much less. If a pressure lower than that set by the forces existing naturally in a liquid is applied, the surface atoms begin to separate and the liquid begins to vapourise at a rate sufficient to balance the pressure difference.

The effect of hydrostatic pressure on a solid is much smaller than on a liquid (temperature being hardly affected at all) and, due to the rigidity of the solid, hydrostatic pressure is not developed from the material's reaction to a unidirectional force; that is the atoms do not so readily slide past each other.

If the volume of a material in the solid state is artificially held constant then local temperature changes can cause marked changes in internal stress, the latter rising and falling with temperature.

Depending on the material and its state, applied pressure can interact on the internal energy balance. Therefore, it is to be expected that applied pressure will affect the temperature at which energy changes are likely to take place. Hence melting temperature and, more particularly, boiling temperature can be changed by changes in ambient pressure. A higher pressure holds atoms more closely together, therefore more thermal energy is needed to force them apart and so the melting and boiling temperatures rise. Rate of vaporisation (see Chapter 3) also varies inversely with pressure for the same reasons.

BIBLIOGRAPHY

MARTIN, J. W., *Elementary Science of Metals*, Chaps 1 and 2, Wykeham, London

MOFFAT, W. G., PEARSALL, G. W., and WULFF, J., *Structure and Properties of Materials*, Vol. 1, Wiley

3 The Makeup of Materials

A particular material's characteristics (see Table 4) are governed by the nature of the individual atoms themselves and by the features of the geometric pattern in which the atoms are disposed. Thus to achieve an understanding of such characteristics certain atomic and structural phenomena must be considered both individually and in relation to each other. These phenomena are:

(1) types of bond
(2) heterogeneity of bonding
(3) molecule formation
(4) crystal formation
(5) agglomeration, aggregation, polycrystallisation
(6) heterogeneity of structure

Each of these is considered in turn in this Chapter.

3.1 TYPES OF BOND

Atoms do not associate with each other by any one simple mechanism. Alternative mechanisms are usually possible and in most cases at least two types will operate simultaneously. A knowledge of the possible types of linking is essential if most of the differences between the physical and chemical properties of different materials and differing states of one material are to be understood.

Such links or bonds are usually subdivided into six types in two main groups of three strong primary bonds and three weak bonds. In the first group there are (1) ionic bonds (2) covalent bonds and (3) metallic bonds. In the second group each of which is a variant of simple van der Waals attractive forces there are (4) molecular polarisation bonds (5) electron dispersion bonds and (6) hydrogen bridge bonds.

3.2 HETEROGENEITY OF BONDING

It has to be clearly understood that no one type of bond operates alone and even the types of bond that do operate will not necessarily operate in the same way under differing conditions within the same material. The obvious illustration of this is the changed behaviour of many materials as they alter in state from solid to liquid and then to vapour with rising temperature.

Naturally, the properties of a material (physical and chemical) are closely related to the types of bonding that prevail within the material under particular conditions. Thus to understand the control of properties requires the understanding of the control of the heterogeneity of bonding.

3.3 MOLECULES

Every material that solidifies in a non crystalline manner is made up of accretions or aggregations of basic stable groups of atoms called molecules. Each molecule is made up of a fixed number of atoms each with a characteristic position within the group and all interrelated to give properties and a geometric configuration characteristic of the material (Figure 3.1). The molecule is often defined as *the*

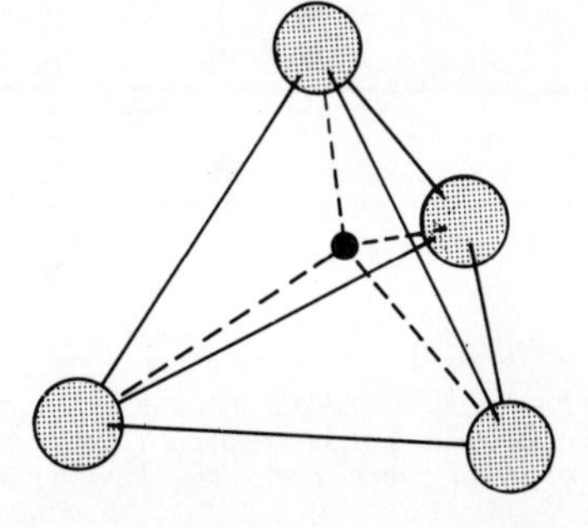

Figure 3.1. Silica molecule (SiO_2) One central silicon atom (black) surrounded by four equally spaced oxygen atoms which can form other bonds with adjacent atoms or molecules See also Figure 5.12

smallest portion of a substance capable of existing independently and retaining the properties of the original substance. Some crystalline substances are also molecular if their structure is made up by the molecules interlinking themselves in the ordered array typical of a crystal.

Molecules of differing substances vary in size, shape and bonding characteristics, so their behaviour patterns when in an aggregate also differ. An understanding of these behaviour patterns is necessary for understanding the bulk properties of any material into which they make up.

3.4 CRYSTALS

Many materials, including all metals, are crystalline in structure. A crystal is formed when the atoms or the basic molecules of a substance order themselves in a simple three-dimensional geometric pattern such that each unit (atom or molecule) is surrounded by exactly the same relative orientation and spacing of immediate neighbours as every other unit within the structure (Figure 3.2).

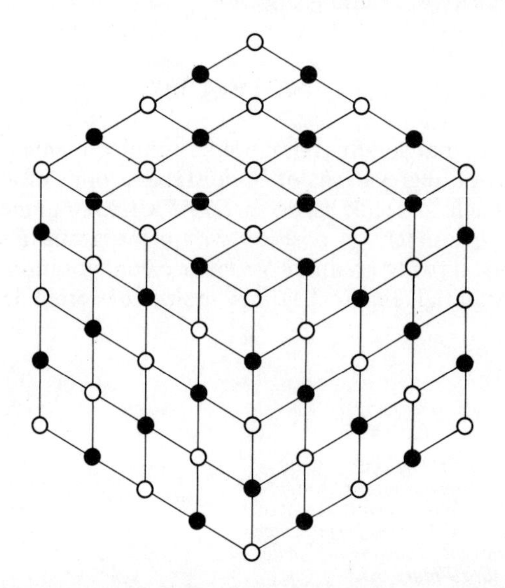

Figure 3.2. Simple orderly array of alternate atoms of sodium (black) and chlorine in sodium chloride (NaCl) with atoms positioned at corners of a stack of equal cubes, each cube forming a unit cell identical with every other cell. See also Figure 3.13

There need be no limit to the number of units going to make up a single crystal which means a crystal does not have a fixed maximum size or shape.

Crystalline materials often possess particular properties because of the order of the crystalline form and, for this reason, the strongest materials are likely to be crystalline. Of course at the surface of a crystal the order has to break down and this may give this region certain properties or weaknesses peculiar to itself.

3.5 AGGLOMERATION AND POLYCRYSTALLISATION

Nearly every material can exist in more than one state. Each state takes its characteristics from the way in which the atoms, molecules or crystals associate with each other to form either (1) a mildly cohesive amorphous volume or fluid or (2) a solid agglomeration of comparatively randomly arranged molecules or (3) a solid agglomreation of small crystals (a polycrystalline structure) or (4) a completely ordered structure made up of one large crystal.

Since only in the least common case of a single crystal is the whole atom structure regular and ordered, it is obvious that the geometric and bonding interrelationships between the molecules, atoms or crystals influence the properties of a particular material.

Where a solid is concerned larger scale structural features and properties are known as the *macrostructural* characteristics in contrast to those of individual molecules and atoms, known as the *micro-structural* characteristics.

3.6 IMPURITIES

It is often impossible to make a material perfectly pure so within its volume there may be quantities of undesirable foreign material which will affect the properties. These impurities may appear as visible isolated groups distributed in the parent material, or they may be finely distributed as thin layers at boundaries between molecules or crystals or they may be spread uniformly throughout the structure. In each case they cause heterogeneity of structure and composition which may have undesirable effects on properties. Of course such heterogeneity may be deliberately caused by adding suitable materials with the definite purpose of adapting properties to some particular application, as is done by the alloying of metals and the hardening of polymers.

3.7 BONDS BETWEEN ATOMS

The point has already been made that most of the differences between the properties of differing materials is explicable on the basis of the types of bonds that may be formed between the atoms within the given materials. However before considering the six basic types of bonds some points must be made to help clarify the complexities of the situation.

In Chapter 2 when describing the situation within a stable atom little attention was given to the possible effects on orbiting electrons of external forces, but in considering bonding, external influences such as electrical and magnetic fields, mechanical forces, thermal effects and particularly the interaction between the atoms must be taken into account.

Completed Principal Quanta are inherently stable and cannot readily be disturbed except by exceptionally strong external forces from appropriate energy sources. Inner Quanta (other than the next-to-outer Principal Quanta of certain atoms) are always completely filled and are particularly stable so they take no apparent part in the bonding process. On the other hand, an incomplete outer Principal Quantum may react to relatively small external forces of a suitable type. As a result, incomplete Principal Quanta play the major role in all the bonding processes.

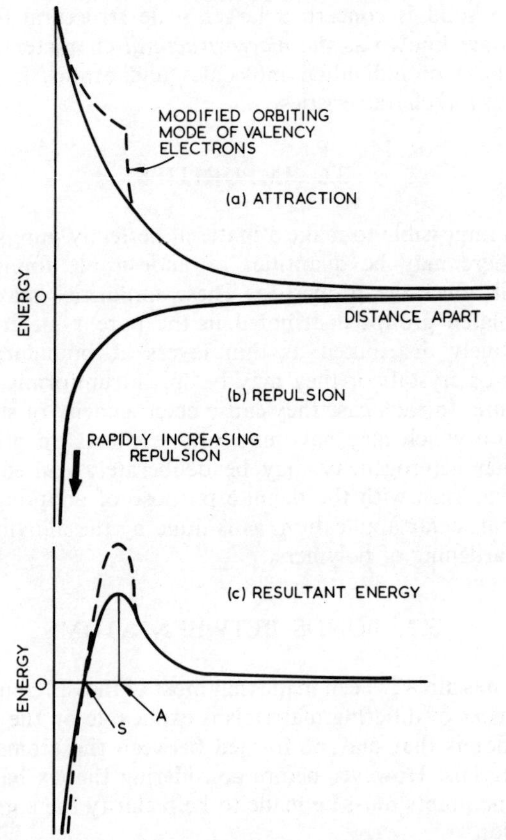

Figure 3.3 Energy balance between a pair of atoms. Variation of energy with distance between centres

When two atoms are gradually brought closer together there may at first be a weak mutually attractive force between them which will tend to increase with decrease in separating distance as shown in Figure 3.3a; but, opposing the attraction, there will begin to build up an increasing repulsive energy between the respective electronegative electrons and a very rapidly increasing energy of mutual repulsion between the respective electropositive nuclear *cores* (a nuclear core may be either a nucleus alone or a nucleus with its associated stable, filled orbitals). The total repulsive energy between the atoms increases rapidly with decreasing distance as shown in Figure 3.3b. When these two energies are summated the energy pattern is on the lines shown in Figure 3.3c. Down to the distance OS the forces are attractive with the maximum at OA; below OS the forces are increasingly repulsive. The atoms will tend to settle at distance OS, if not influenced by external energy forces, since externally-applied force is required to move them in either direction from that distance. The distance OS depends on the distance at which the magnitudes of the attractive and repulsive energies balance and the firmness with which it is held in that position depends on the slope of the curve. The steeper the slope (i.e. the higher the rate of increase of energy) the more rigidly the distance will be maintained. The maximum force required for total separation will be proportionate to the height of the peak of the curve.

In certain circumstances the attractive energy between atoms may show a sharp increase (see broken line in Figure 3.3a) with decreasing separation. Such circumstances arise, for example when a relative rearrangement of outer orbitals between the two atoms makes the total orbital system more stable. In such a case, as the two atoms approach each other, the first effect is a slight displacement and perhaps some distortion of the orbitals (Figures 3.4a and b) as the

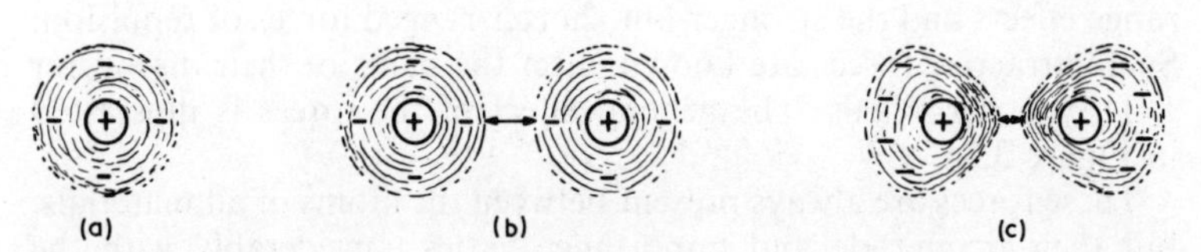

Figure 3.4. Reaction between spherical orbitals of adjacent atoms as they move closer together. (a) Undisturbed spherical orbital. Electrical potential zero. (b) Spherical orbitals slightly repelled, potential hardly affected .(c) Valency orbitals strongly repelled with marked distortion giving polarisation of charges

valency electrons repel each other. Closer approach causes much more severe distortion (Figure 3.4c) associated with some broadening in the range of energy tolerance level for electrons within each quantum, and allows the mutual repulsion of the nuclear cores to increase since they are no longer so effectively shrouded (i.e. the atoms are tending to become ionised and are more electro-positive on their near sides and more electronegative on their far sides, see Figure 3.4c). At this stage, due to mutual repulsion between respectively orbiting valency electrons reacting with their simultaneous attractions to the opposed slightly unbalanced cores, mutually resonant electron movement may be set up across the boundary between the atoms. This may occur by resonant pairing of a vibrating valency electron in one atom with another in the second atom made easier by the broadening of the energy bands. As a result a significant local attraction may be set up between the two atom assemblies drawing them closer together and holding them more firmly at a more stable distance (see broken line Figure 3.3c). It may even be possible for electrons to be shared between the nuclear cores in a new, joint, stable orbital system giving great stability to the combined system.

3.8 VAN DER WAALS' FORCES

It was shown in Chapter 2 that all the electrons in the systems of an atom are in continuous seemingly random movement within their orbitals with the result that the negative electropotential of an atom appears to pulsate, particularly with respect to the valency electrons.

Owing to this pulsation of the negative potential there is a resultant long-range weak attraction operating between each respective nucleus and each electron group of every other adjacent atom. These attractions tend to draw atoms closer together and may hold them quite firmly at the position of balance between the weak longer-range effects and the stronger but shorter-ranged forces of repulsion. Such attractive forces are known, after the name of their discoverer J. D. van der Waals. The general effect of the forces is illustrated in Figure 3.3a.

These forces are always present between the atoms of all materials, but their magnitude and importance varies considerably with the type of material. Their lowest values are found between those atoms with completely filled Principal Quanta and their highest values are likely between atoms of low valency. Normally the forces with which they hold atoms together are not great, therefore, materials in which they are the principal bonding agency are not likely to be mechanically strong.

In general, attractive forces of this kind are not strongly directional relative to the angular orientation of the atoms concerned but, in certain circumstances directional tendencies do appear and the local forces of attraction can then be so increased that their effects on mechanical strength become significant, particularly if this type of force is the main linking mechanism.

3.9 IONIC BONDING

Outer shell or valency electron stability of an atom depends on the relative number of electrons normally present in that Quantum. Except for hydrogen and helium the valency PQN of each atom has a maximum valency capacity of 8 electrons for best orbital stability. If only a few electrons are normally present it is easy to unbalance their movement (Figure 3.4c) or even strip them off and leave the atom *positively ionised* (see Figure 3.5). In an atom with a

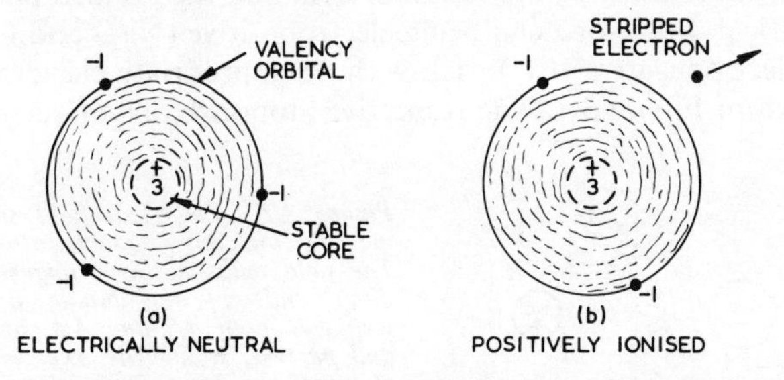

Figure 3.5. Atom with low number of valency electrons losing one electron left positively ionised

high number of valency electrons the valency orbitals will not move so easily and it will be much more difficult to strip off an electron; but it becomes easier to capture stray electrons to complete the shell, leaving the atom *negatively ionised* as shown in Figure 3.6.

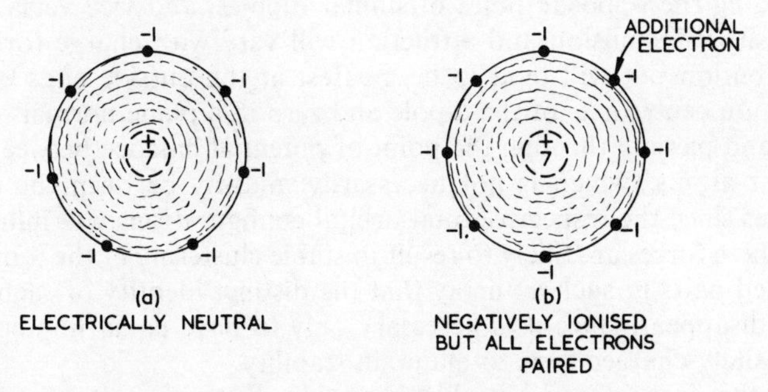

Figure 3.6. Atom with incomplete valency shell captures an electron but becomes negatively ionised

It is obvious that if these two possibilities exist, two types of atom, one with few valency electrons and another short by the same number might have strong affinities for each other if brought close together. Indeed this forms the basis of the ionic bond.

The best-known illustration of this bond is common table salt in which the basic sodium chloride (NaCl) molecule is formed by a sodium (Na) atom giving up its one valency electron to a chlorine (Cl) atom which normally has seven. Since the sodium atom is then positively charged (Na^+) and the chlorine atom is negatively charged (Cl^-) the two will be strongly drawn to each other by coulombic attraction and will hold firmly together at a centre distance determined by the balancing (minimum energy) position between the forces of coulombic attraction and the forces of nuclear repulsion. Although the nominal electrical potential of such a pair of atoms is zero, the mutual ionisation is still present and the bonded pair form an electrical dipole one end being electropositive (Na^+) and one end being electronegative (Cl^-) each with an appropriate electrical field as shown in Figure 3.7. The respective atom orbitals of such a pair

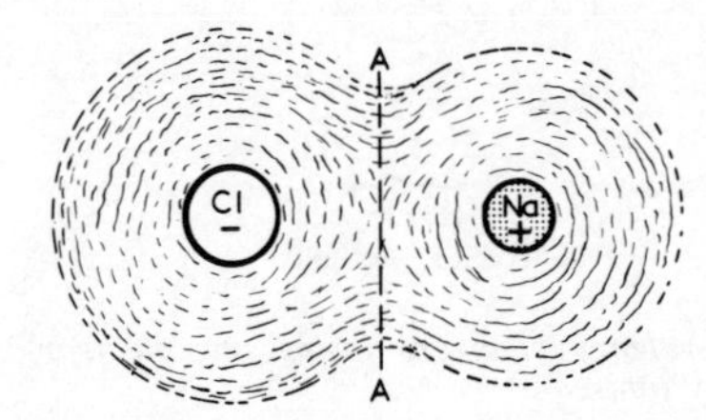

Figure 3.7. Electrical field surrounding an ionically bonded atom pair. The field tends to be strongest on the outer ends of the dipole axis and weakest about a plane AA, normal to and partway along the axis where the charges balance. Thus the opposite ends of adjacent pairs would attract and similar ends repel

may not be of the simple shape suggested and their basic geometry will be subject to distortion from mutual interorbital repulsion; but the poles will certainly be attractive to charges of opposite sign, including the opposite poles of similar dipoles, and vice versa. The intensity of repulsion and attraction will vary with charge (orbital) distribution but will tend to be greatest at the outside ends of the common centre axis of the dipole and zero in a plane normal to the axis and passing through the point of potential balance between the parent atoms (which is not necessarily midway between the atom centres since the respective total orbital configurations will influence it). These forces are likely to result in stable clustering of the ionically bonded pairs in such an array that the distinct identity of each pair may disappear, since it is necessary only to have equal numbers of oppositely-charged ions to maintain stability.

In the case of sodium chloride such clustering takes a simple orderly form and the atoms group themselves as shown in Figure 3.8.

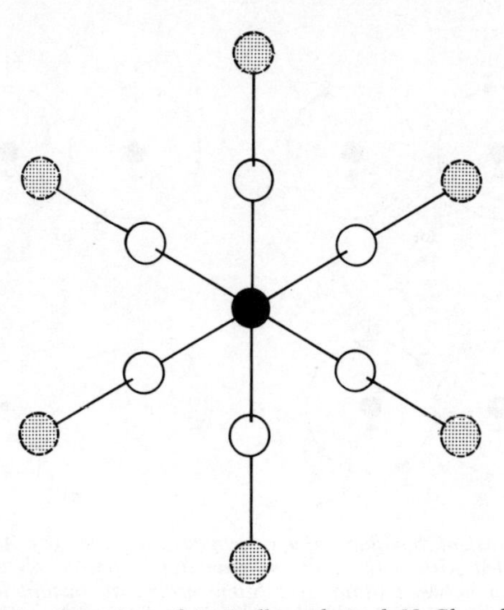

Figure 3.8. Basic clustering of ionically polarised NaCl molecules. Six Cl atoms (light circles) equally adjacent to one Na atom

so that each atom of one kind is closely surrounded by 6 equidistant atoms of the opposite kind, equally spaced from each other, and by 6 similar atoms also equidistant from the reference atom but $\sqrt{2} \times$ the first mentioned distance from it and from each other. This gives what is known as a *crystalline* structure to the solid, of a form known as cubic. The spacing distances and the pattern are both determined by the balance between all the relevant mutually active forces. This particular form is not the only ionic bond configuration possible. Other crystalline patterns are possible depending on the number of close neighbours that each atom will tolerate (coordination number) and on relative sizes etc. or an irregular clustering may result, particularly if ionic bonding is not the only principal bonding mechanism or if more than two types of atom are interacting.

3.10 COVALENT BONDING

Covalent bonding, alternatively called homopolar bonding, which can be very strong, is possible because of the mobility of valency electrons in an incomplete outer shell relative to the stability of a completely filled outer shell. If two suitable atoms are brought so near to each other that they can interchange two or more valency electrons and share them to give alternately filled outer shells the atoms become strongly attracted to each other (see Figure 3.9).

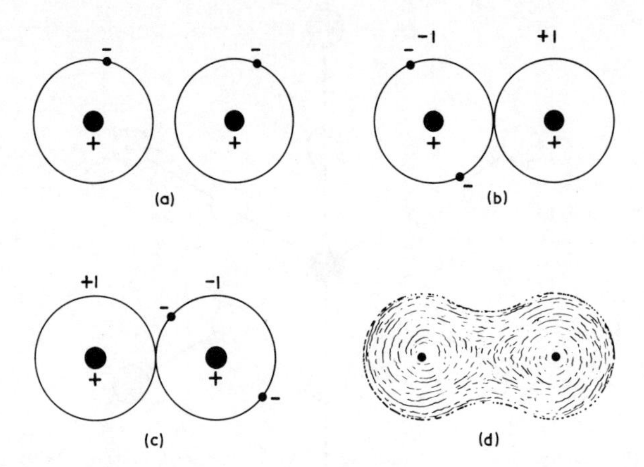

Figure 3.9. Covalent bonding in a hydrogen molecule. (a) Atoms separate. (b) Atoms coupled with both electrons transitorily paired about one nucleus. Strong attraction between atoms. (c) Same atoms an instant later. Electrons paired about second nucleus. (d) Mutual orbital formed by hydrogen molecule

Covalent bonding can occur between both similar and widely differing pairs of atoms and may occur simultaneously between one atom and two or more others to build up a complex molecule or a larger aggregation. The bond is formed because an atom which is in the state of borrowing an electron becomes electronegative and the atom which is lending an electron becomes electropositive and the two are mutually attracted. When the situation reverses the attraction remains because the potentials have reversed.

A very simple example of covalent bonding is the hydrogen molecule shown in Figure 3.9. A PQN1 shell is completely filled with 2 electrons but the hydrogen atom has only one electron. If however, hydrogen atoms are brought close together they will under normal conditions pair off, so that each pair shares their two electrons in such a way that their valency shells are alternately full, each atom being strongly attracted to the other in the process. In this state the independent orbitals join and change from two separate spherical orbitals to become one dumbbell-shaped orbital with the ends of the dumbbell approximately centred on the respective nuclei. Since all other atoms, except helium, require 8 electrons to complete the normal valency shell this kind of sharing must involve either the intermittent provision of 8 electrons on each side of the bond or hydrogen atoms may be included in the system to give a 2 to 8 configuration by the mutual resonant sharing of 2 electrons. If an atom (except hydrogen) is to form a stable covalent bond with its own kind it must have at least four valency electrons if sufficient stability is to be given to the covalent association. Hence, atoms with less than four valency electrons each (except hydrogen) are unable to form a simple stable covalent system, but atoms with four or more may be able to form such a stable association.

A typical example is the oxygen molecule (oxygen valency = 6). In this molecule four electrons, two from each atom, are always shared mutually in such a way that if, at one given instant one atom has four valency electrons closely associated with it, its partner has eight closely associated with it (see Figure 3.10) and in the next

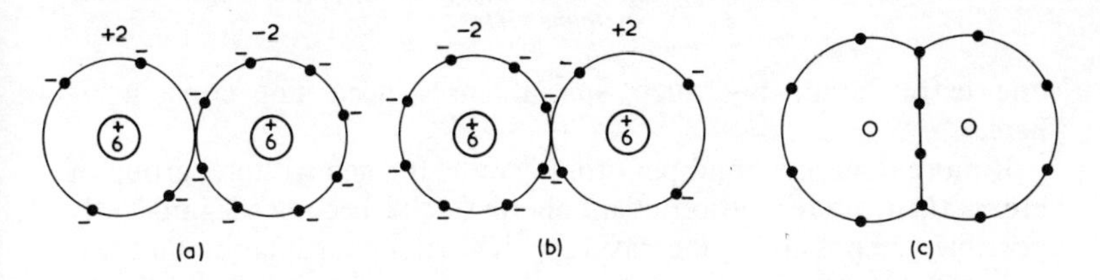

Figure 3.10. Covalency in an oxygen molecule. (a) Pair sharing 4–8. (b) Instant later pair sharing 8–4. (c) Conventional representation. Nuclear centre distance may be given

instant the associations are eight and four respectively. The electrical charges of each atom in the first case are positive and negative and are reversed in the second state so there is always a strong mutual attraction between the atoms. Since the electrical dipolar condition of a covalent bonded pair is in a state of continual reversal, the electro-potential fields associated with it tend to average each other out in contrast to the effect of the relatively stable polarised condition of an ionically-bonded pair. As a result there is not normally any *steady* external coulombic attraction or repulsion associated with a covalent bond.

Figure 3.10c is a conventional way of indicating a covalent relationship between atoms. In this system each dot represents one valency electron and the identity of an atom is indicated by its element symbol at its imaginary centre (*O* in this case).

Although covalent bonding has been considered only between identical-type atoms, it is not limited in this respect and suitable dissimilar atoms may bond in this way, see Figure 3.11. Furthermore, suitable atoms may be able to form more than one covalent association simultaneously. The maximum number of possible associations for such atoms (the *coordination number*) being, usually, limited to 8 − N, where N is the valency number of the atom. (There are some exceptions to this rule, notably where an atom for some special reason is able to contribute two of its valency electrons to

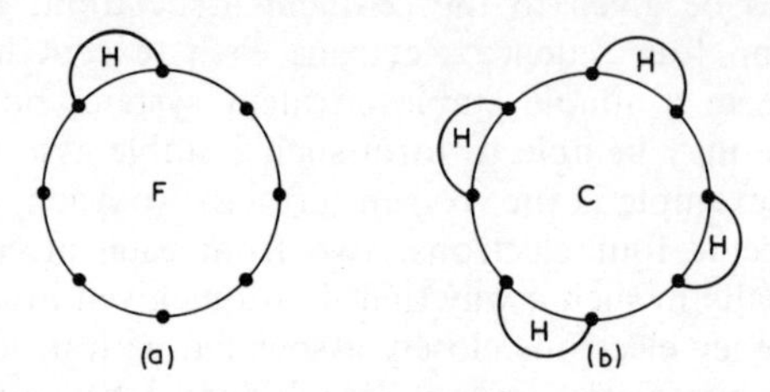

Figure 3.11. Covalent compound molecules representation. (a) Fluorine ($N = 7$) plus hydrogen ($N = 1$. (b) Carbon ($N = 4$) and hydrogen forming methane CH_4

one extra bond, but such special cases need not concern us here.)

If more than one bond per atom can be formed within a group of atoms then the formation of a coherent solid becomes significantly possible, the possibility increasing with greater coordination number. A possibility of three bonds per atom makes a strong three-dimensional network attainable, and the strongest solid material known, namely diamond, derives its strength from the ordered four-link covalent bonding of each constituent carbon atom. On the other hand, the mere fact that numerous strong covalent bonds have formed within a solid does not make it certain that the solid as a whole will be inherently strong. The latter will depend on the nature of the bonds, the relative number of bonds actually formed and arrangements of the bond within the material. One other statement can be made, namely that the more valency electrons there are mutually shared in an individual bond, the stronger that individual bond is likely to be (because the transient electropotential differences are greater). On the other hand, the fewer are the chances of a second strong bond forming with either one of the two atoms forming the first bond.

Before leaving covalent bonding it is worth thinking again about the particular nature of the hydrogen atom. Because of its valency state (N=1 and filled state 2) a hydrogen atom can form one covalent bond with any one of a very wide variety of other atoms. Also, more than one hydrogen atom may bond to one other atom of a suitable kind perhaps in conjunction with a third kind of atom or atoms. The outstanding association of hydrogen in this way is with the carbon atom with which it can associate in conjunction with other atoms to form a wide range of *hydrocarbon* molecules. Thus, we may find in one example of the first situation a single hydrogen atom locked to a single fluorine atom (N = 7) to form a molecule of hydrogen fluoride, HF (Figure 3.11a) and in an example of the second situation we may find four hydrogen atoms locked on to a carbon atom to form a molecule of methane, CH_4, as shown in Figure 3.11b.

3.11 METALLIC BONDING

In each of the bonds described so far the main bonding forces have arisen from electropotential differences between adjacent atoms, created by the mobility of the electro-negative valency electrons as they react with each other, with their own, and with adjacent electro-positive nuclear cores. These mobility effects are carried to their limit in the metallic bond in which all the valency electrons of all the participating atoms are mutually shared in a complex joint orbital system. Some workers regard this situation only as a particular example of valency bonding but it has a number of features peculiar to itself and is the principal bonding system formed in the materials included in the range known as metals.

It is impossible to visualise the system of bonding but it is commonly described as an electro-negative cloud of electrons enveloping and holding together a group of electro-positively ionised atoms by means of the forces of mutual attraction of the cloud and group for each other. The respective mutual repulsions between electrons and between ionised cores is overcome by the multiple interactions within the system. It is clear that, because (a) a relatively large minimum number of atoms is needed to create conditions suitable for such bonding to stabilise and (b) the minimum number varies with ambient conditions, a basic molecule of substance bonding in this way cannot be defined in the way that molecules of most other substances can. Equally, there is no theoretical upper limit to the size of a single metallically-bonded group of this kind, so a molecule cannot be defined in terms of an upper limit of numbers of constituent atoms. This is not to suggest that such a substance is physically and chemically undefinable but that it must be identified by other means.

Metallic bonding within a solid results in a distribution of the constituent atomic cores into a uniform, geometrically-ordered pattern of positioning relative to each other; that is, they form a *crystalline* structure. This atom positioning occurs as a direct outcome of the uniform coulombic forces, operating around each ionised core, interacting in such a way that each ionised atom is surrounded by exactly the same number of ionised neighbours, each located in a similar relative position and at a similar relative distance so that the combined attractive and repulsive forces are in a state of statistical balance. A material held together in this way can be identified by the characteristic type and proportions of its crystalline structure.

What determines the crystalline pattern of arrangement into which a metallically bonded solid material will form is the maximum coordination number of the principal atoms of which the structure is made up. The coordination number is itself determined by (1) the respective numbers of valency electrons in each atom (2) the effective size of the ionized core of each atom and (3) the final configurations of the shared orbitals making up the system. Ambient conditions of temperature and/or pressure can also systematically influence the characteristic pattern of atom distribution within some metallically bonded materials.

To form a metallic bond the participating atoms must have relatively mobile valency electrons in orbitals which, when the atoms are brought close together overlap and intersect with each other so that a valency electron can readily transfer from an orbital centred on one atom into an intersecting orbital of an adjacent atom and thence into a similar intersecting orbital of another atom and so on. All the available valency electrons present share in the movement and travel at high speed in and about the common orbital paths available between the atom cores. It is here that the cloud analogy fails to give a true picture of the valency electron state. These electrons do not move in completely random paths but keep to a closely interlinked three-dimensional orbital network of easy paths, making the common valency orbital system seem more like a sponge network than a cloud, with atom cores locating themselves in appropriate interstices of the sponge orbitals. Different sponge arrays, depending on the basic orbital configurations of the parent atoms, will form with differing parent atoms so that the relative freedom of valency electron movement will vary between materials and may vary with direction within a given material.

Since valency electrons are free to travel within the common orbital system in a metallically bonded material, two phenomena are characteristic of such a material.

(1) It will be relatively electrically conductive. That is, if an electric power source with a suitable potential difference is put in series with the material a stream of electrons will pass through the material by way of the common orbital system.

(2) It will be a relatively good thermal conductor. That is, if the atoms at one part of a metallic material are thermally agitated their agitation will readily transfer itself through the material by increased rate of exchange of valency electrons between the parts (see Chapter 4).

Some dissimilar metal atoms may readily form metallic bonds and others which may not be so ready to bond in that way may bond by other mechanisms. Because a material is classed as a metal it does not mean that metallic bonding is the only way in which it can bond and a metal atom may readily bond by other means to a suitable non metallic atom or atoms. In the latter case the characteristic properties of the metal atom may be lost in the union.

3.12 MOLECULAR POLARISATION BONDING

In Section 3.9 the formation of a relatively stable molecular electrical dipole as a result of ionic bonding between pairs of suitable atoms was shown to give a mutual attraction effect between the dissimilar poles of adjacent molecules, thus enabling an ordered array or group of molecules to bind firmly to each other in suitable circumstances. This effect, called molecular polarisation bonding, is not confined to ionically bonded pairs but may also occur with some valency bonded pairs. The bonding forces in the latter case may be relatively weaker because the electropotential difference is likely to be lower.

In valency bonding a dipole may exist when the bonded atoms have widely differing core sizes (e.g. when a hydrogen atom forms a valency bond with a larger atom). The nucleus of the atom with the lower total number of electrons of the pair will be, on average, less well shielded by its orbitals than will the nucleus with the larger number of electrons. Therefore the former atom will seem to be slightly electropositive and the latter slightly electronegative with respect to outside neighbours. If opposed dipoles can settle into a reasonably balanced relationship by suitable positioning of adjacent molecules, then a stable group may build up although its bonding energy is likely to be low and the possible range of order in the group limited.

3.13 ELECTRON DISPERSION BONDS

Since the electron orbitals surrounding an atom give a pulsating electronegative field then all atoms will have frequent slight momentary electropotential differences existing between them, regardless of any other forces that operate between them. As a result of these effects there will always tend to be a very slight mutually attractive force between atoms and molecules. These attractions are so slight that they play a negligible part in the bonding of most materials and become significant only between those atoms or molecules which have no other mutual bonding mechanism.

Because these forces are a result of the distribution pattern of electrons related to the nuclei of atoms they are called dispersion bonds.

3.14 HYDROGEN BRIDGE BONDS

The last significant type of bond occurs as a result of the kind of molecule formed when two similar small atoms form valency bonds with one larger atom, notably as with hydrogen bonded to oxygen (H_2O) in water. Because of this it is known as the *hydrogen bridge* bond.

This bonding is a molecular polarisation effect due to poor screening of the two valency-bonded smaller atoms relative to the surface of the single high-valency atom. In the isolated state, a molecule of this type would tend to be symmetrical as in Figure 3.12a with the

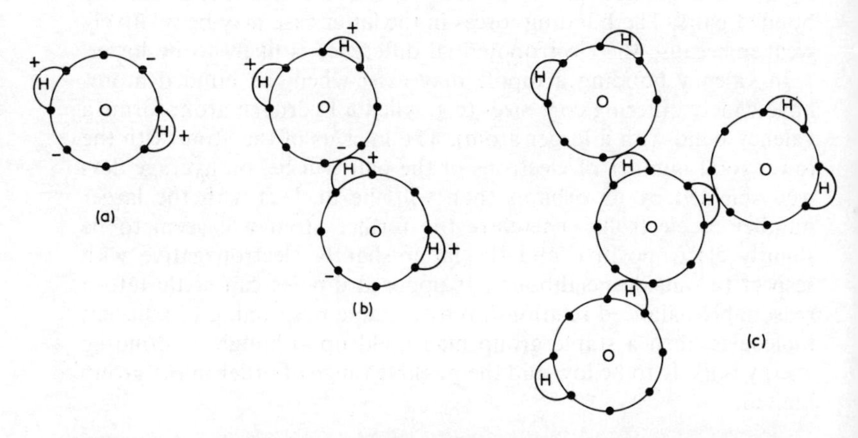

Figure 3.12. Hydrogen bridge bonds between water molecules. (a) An undisturbed molecule in approximate electrical balance. (b) Molecules forming dipoles and linking to each other. (c) Molecules linking into a 3-dimensional network

positively ionised atoms (hydrogen in this case) tending to be diametrically opposite to each other so that their electrical energies balance. Now if a second molecule is brought near to the first (Figure 3.12b) the slightly electronegative group of uncombined valency electrons of the large atom (oxygen in this case) will be attractive to the more positive hydrogen ions of the second molecule. Therefore one of the latter will be attracted to the first oxygen atom, repelling the already attached hydrogen atoms which are misplaced enough to form a linking dipole in the first molecule. The second molecule will also tend to form a similar polarised system (both its hydrogen ions are attracted) and could be further attracted to a third molecule (Figure 3.12c) and so on, making a three-dimensional chain network possible (as in ice) since the linking is not limited to the two-dimensional orientation of the diagram.

Such bonds are not very strong but are much stronger than dispersion bonds in those materials which they operate.

3.15 CHEMICAL PROPERTIES AND BONDING

Because of the consistent characteristics of the atomic configuration in a cluster of atoms of any one type, it is obvious that such atoms will react with each other in certain simple limited ways. Equally a group of the same atoms can be expected to behave in different, but predictable ways in relation to the presence of atoms of a different type. These differences in behaviour, called the chemical properties of materials are the direct result of bonding reactions between the atoms and/or molecules.

Atoms with all their Principal Quanta completely filled are inherently stable, therefore atoms of this type are chemically inert and unlikely to react either between themselves or with any other types of atoms. Thus helium (N=2 in PQN1) and neon (N=8 in PQN2) are elements which are normally monatomic gases. Their atoms do not combine either with their own kind or with other elements with any strength. In fact the only potential bonding mechanism is dispersion bonding and this can be effective only in the almost complete absence of thermal excitation. That is they condense into a liquid only at very low temperatures, 4·1 and 28·1K respectively (0°C = 273K) and solidify at still lower temperatures. Similarly, atoms with eight valency electrons (N = 8) even although the valency PQN is not completely filled, are stable and unreactive. Thus argon, krypton, zenon and radon are all normally found as monatomic gases and liquify only at temperatures below 87·3, 120·1, 165·9 and 211·2K, respectively. The fact that these temperatures are significantly higher than the previous two shows that their respective dispersion bonding forces are correspondingly greater.

Atoms of all other less stable elements will be more chemically active and will react in at least one other way with their own kind or with other kinds of atoms. Within certain limits, the less stable its inherent atomic configuration the more likely is a particular atom to be chemically active and to react in different ways and with differing atoms. Thus low-strength covalent bonds will form spontaneously within isolated finite groups of atoms of elements such as hydrogen or oxygen which, at n.t.p. always form into diatomic molecules, relatively more stable in this paired grouping than in the atomic state. Dissimilar atoms may bond in similar or perhaps stronger and more complex ways to form compound molecules. Molecular stability is relative, therefore if molecules of one kind are mixed with another kind to which their constituent atoms can bond more strongly, and the change is initiated, then there will be a recombination of atoms into a new form of molecular compound. Thus, if oxygen (O_2) is suitably mixed with hydrogen (H_2) and a flame is applied, the two will combine to form water molecules (H_2O).

Similar molecules may react in certain limited ways with each other, each molecule acting as a unit rather like an atom. Thus hydrogen gas condenses into liquid if its temperature is lowered below 21K; oxygen gas condenses into liquid if its temperature drops below 90K, water molecules condense into liquid if the temperature is dropped below 373K, the linking mechanism being mainly electron dispersion bonding. In the case of the water molecules, if temperature is lowered below 273K a further reaction takes place, by the added formation of hydrogen bridge bonds, and the water becomes ice.

Specific amounts of heat energy (*latent heat*) are released as atoms or molecules rearrange themselves into more stable arrays and similar amounts of extra heat must be absorbed before the process can reverse, the amounts depending on the kind of atom or molecule and the nature of the change in relation to the quantum mechanics of the system. Chemical changes vary greatly in their speed and manner of change. Combustion in which O_2 molecules combine rapidly with atoms or molecules of a combustible substance to form an oxide compound, is one of the best known chemical recombination processes. This process is potentially violent but it may occur quite gently and almost imperceptably as slow oxidation at low temperatures.

Molecules of a compound may break down in a chemical reaction by giving up only one of the types of atoms they contain, to form a new compound and leave the remaining types of atoms either as a separated element or, if more than one type is left, possibly as an alternative, simpler, compound. Such reactions are well known in ordinary chemical practice and form the basis of nearly all chemical treatments and processes. The relative ease or difficulty of starting and propagating a chemical reaction depends on (1) the various types of bonds operating or potentially operative within and between the substances under consideration (2) proximity to each other of the particles of the substances (influenced by particle size, particle shape, degree of mixing, type of contact relationship and environment) (3) availability of sufficient energy of an appropriate type to initiate the interchanges between the atoms and (4) sufficient proportions of the substances being present to maintain the reaction once it begins. Usually, some extra energy difference is required at the beginning of a reaction to start it off, then it may become self propagating or settle down at a rate depending on the nature of the reaction and the resulting compound. A reaction may be induced and/or speeded up by means appropriate to the circumstances. Change in thermal conditions, already mentioned in relation to melting, freezing, boiling and condensation, is one of the most common of these means, but it is not the only one. Applied pressure may change reaction conditions and can sometimes initiate a reaction by bringing the relevant atoms closer together into the operation range of previously ineffective short-range energy forces. The effects of pressure may also depend on the rate at which the pressure is applied, as is shown by the need for a small easily exploded relatively unsafe detonator to initiate controlled explosion of the large controlled bulk of an industrial or military explosive charge which has to be suitable for safe storing and handling in bulk before use. In a similar context, the relative explosive sensitivity of the liquid nitro-glycerine ($C_3H_5(NO_3)_3$) which may explode if given a relatively slight impact shock, compared with dynamite (liquid nitro-glycerine absorbed into a porous substance to convert the mixture into a stiff paste) which needs a heavy impact to detonate it, is an illustration of the effect of proximity of molecules on the initiating energy needed for reaction. Other means of initiation may include ionisation (e.g. by electrolysis or in an electric arc), irradiation, catalysis and increased surface contact (stirring).

The chemical activity of a substance may be influenced by change in the geometric distribution of identical types and numbers of bonds within that substance. Many substances are *polymorphous*, that is may contain the same proportions of constituent atoms or molecules but may have them bonded or arranged in different ways in different circumstances. The chemical influence of such differences depends on the relative stabilities of the respective arrangements in each situation. In some cases there is no influence, as in the case of truly allotropic change in which changed bond arrangement changes the physical nature of the substance but does not affect its chemical nature. On the other hand chemical reaction does change with differing bond arrays in other substances. Carbon is an example of the latter; as diamond, carbon is much less reactive than when it is in its graphite form (atoms combined into loosely interconnected sheets or layers) which in turn is less reactive than the charcoal form (atoms heterogeneously arranged).

As one would expect, compounds have very different properties from their individual parent elements. For example, an oxide by its nature is more refractory than the original combustible element from which it formed, its melting point will differ and it may react with other substances to which its parent element was resistant and vice versa. A simple example is the combination of aluminium (a metal, m.p. 932K) with oxygen (diatomic gas, m.p. 54K) to form alumina (Al_2O_3, m.p. 2270K). Alumina is one of the most refractory of common materials and may exist as a soft amorphous powder easily soluble in acids, but by heating to a sufficient temperature it may be bonded into an acid-resistant, hard, refractory substance or by melting and resolidification may be crystallised into a hard transparent abrasive corundum (basically the same as ruby and sapphire jewels).

3.16 MOLECULES

A rough definition of a molecule has already been given but it is now necessary to have a clearer picture of their characteristics and limitations.

If a minimum of either one type of atom, or of fixed proportions of two or more types of atoms (a compound), is able to exist as a stable integrated independent group then the group constitutes a molecule. Each molecule will have the same chemical and physical properties as every other molecule as long as its internal group arrangement remains the same. The essential feature is that the molecular unit is more firmly bonded within itself than is either one molecule with another or any other configuration. The molecule tends to behave as a unit block or *monomer* from which a material may be made up. Therefore, if the behaviour patterns of one molecule are known its potentialities for bonding with its own kind or other kinds of molecules or with independent atoms are predictable.

Molecules may associate with each other in a stable geometric array called a *polymer*, a stable grouping of monomers, held together principally by bonds of the molecular polarisation, the electron dispersion and/or the hydrogen-bridge type according to the nature of the internal bonds and the nature of the constituent atoms. Alternatively, if such external bonds do not form readily between molecules, it may be possible to interlink the molecules by forming intermediate or bridge, bonds by means of a specially-added type of atom or molecule, each of which must be able to bond to at least two of the parent molecules at one time and so build up a polymer. Such interlinking may be based on a general attraction between each interlinking agent and each relevant molecule or on a particular local polarisation attraction between the link and a specific part of the parent molecule as in the hydrogen-bridge type of bond.

External bonds between molecules may not be as strong individually as the basic types of internal bonds inside the molecule, but the variety of possibilities in respect to geometric arrangement, particularly with linking agents, is likely to be wide and may open up a whole range of arrangement possibilities from random orientation systems, through clusters, chains, coils and rings, to geometrically repetitive crystal types of array based on the unit molecule (see Chapter 5).

Molecular materials tend to behave in the normal way with respect to temperature and are likely to have characteristic melting and boiling temperatures. However, if the molecules are of a compound type, anomalies may appear, melting and boiling may not occur as simple processes at fixed temperature but may spread over a range of temperature or may even cease to be clearly distinguishable at all, or possibly may be superseded by irreversible recombinations of the constituent atoms if the temperature is raised to a critical level.

3.17 CRYSTALS

If the atoms or molecules of a substance are to form into a crystal array they must conform to certain rules of arrangement, such that each constituent atom or molecule is surrounded by the same pattern and orientation of neighbouring atoms as every other similarly positioned atom or molecule and hence, must be in a specific position relative to an imaginary three-dimensional framework of lines called a *space lattice*. A space lattice is made up of three sets of intersecting parallel lines, each set being equally spaced within itself in such a way that every single intersection point between a pair of lines is also an intersecting point for a line from the third set so that the space is made up of a stack of parallelopipeds of identical size, shape, and orientation on the lines shown in Figure 3.13. Although there are countless varieties of crystals, there can be only 14 types of space lattice. Each crystal lattice may be divided into *unit cells*, each identical and giving the minimum configuration of atoms and/or molecules typical of the material as it exists in that state. A unit cell may be made up of one parallelopiped of the particular lattice or it may embrace a suitably symmetrical group, but within its volume it

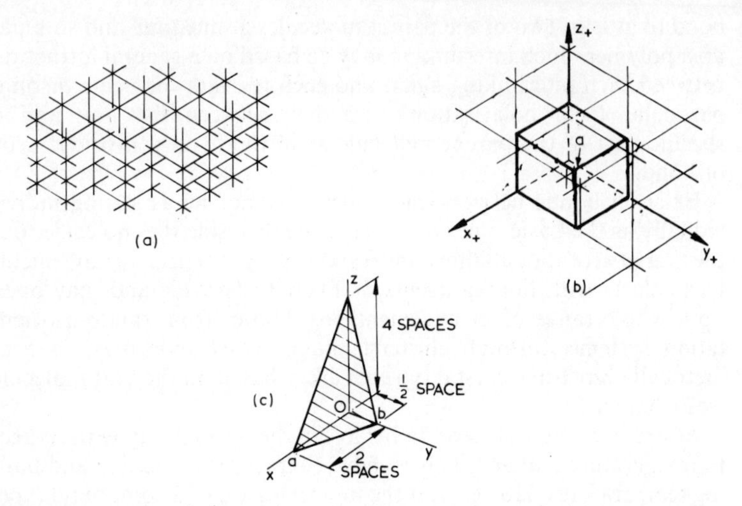

Figure 3.13. Use of a space lattice in the study of crystals. (a) 3-axis system. (b) A unit cell with reference axes. (c) Crystallographic plane intersecting typical cell axes

must contain the minimum number of atoms or molecules and the exact proportion of types of atoms or molecules, typical of the material, located in the same relative orientations and positions as in every other unit cell. Atoms or molecules lying on the outer surface of a cell are shared with the adjacent cell or cells (an edge position is shared between four cells and a corner position between eight cells). A unit cell may be regarded as a kind of molecule of the substance but it cannot exist by itself, nor can it constitute a crystal on its own.

If a crystal structure is to be stable it must contain a considerable number of properly stacked unit cells, but there is no specific minimum number, because that number will vary with ambient conditions. On the other hand there is no theoretical maximum to the number of cells in a particular type of crystal. In a finite crystal, the outer cells are not completely surrounded by similar cells, therefore they are not so firmly located as inner cells and are likely to react with other crystals or substances with which they can come into contact, distorting, detaching, disintegrating or absorbing foreign atoms or molecules according to the nature of the possible reactions and the prevailing conditions at the surface.

The type of crystal array that forms within a substance depends on (1) the types of bonding forces which prevail (2) the stability of the orbital systems of its parent atoms and/or molecules and (3) the ionization patterns that develop within the orbitals of its atoms or molecules; but it must be possible for the constituent atoms and/or molecules to make up a unit cell and be capable of bonding to six neighbouring cells, of the same basic pattern as the first cell and suitably disposed about it. The natures and strengths of the bond system that holds one cell to another need not be identical on each side of each cell but all opposing sides must have identical systems if the symmetry of the lattice is to be retained. Bonding differences between adjacent sides are particularly likely with crystal structures containing molecular-unit constituents.

A unit cell is usually described in terms of three imaginary reference axes originating at its back, left-hand, bottom corner and following its edges, see Figure 3.13b. The unit of dimension used on each axis direction is the size of the cell *on that axis*; thus, the unit will differ on each axis if the true sizes of the cell differ on the respective axes. Note that the axes are at the same angles to each other as are the edges of the cell and only with orthorhombic and cubic types of cell will they be mutually at 90°. A position in a lattice is located, relative to the reference cell, by quoting its three ordinates, on the respective axes, enclosed in double square brackets. Position *a* in Figure 13b is [[1,1,1]] whilst a direction is given by quoting within single square brackets the ordinates of a point through which an appropriate arrow, starting from the origin, would pass. If the latter reference was meant to apply to any similar type of direction within a lattice (a lattice need not be viewed from one particular direction and completely symmetrical lattices have several similar alternative viewing directions) then the ordinates will be enclosed in angular brackets. Negative directions are indicated by putting a bar above the figure. One characteristic feature of a lattice is that the constituent units (atoms or molecules) can be seen to be distributed in regular positions within particular planes of distribution within the lattice to form *crystallographic planes* (any crystal system can be visualised as completely made up of stacks of similar parallel planes suitably piled one on top of the other, there being many alternative ways of viewing the make-up of such planes). It is often necessary to indicate the orientation of particular types of planes and this indication is achieved by using the relevant *Miller indices* and enclosing them in simple parenthesis. Miller indices are obtained by finding each intersecting distance of a typical plane along each particular axis of a system. In Figure 3.13c, Oa = 2, Ob = $\frac{1}{2}$ and Oc = 4, finding the inverses of these distances ($\frac{1}{2}$, 2, $\frac{1}{4}$) and converting them to the lowest integer set (2, 8, 1). A Miller index gives the relative frequency of intersection of the relevant plane system along the respective axis.

A general understanding of the principles of these systems is helpful

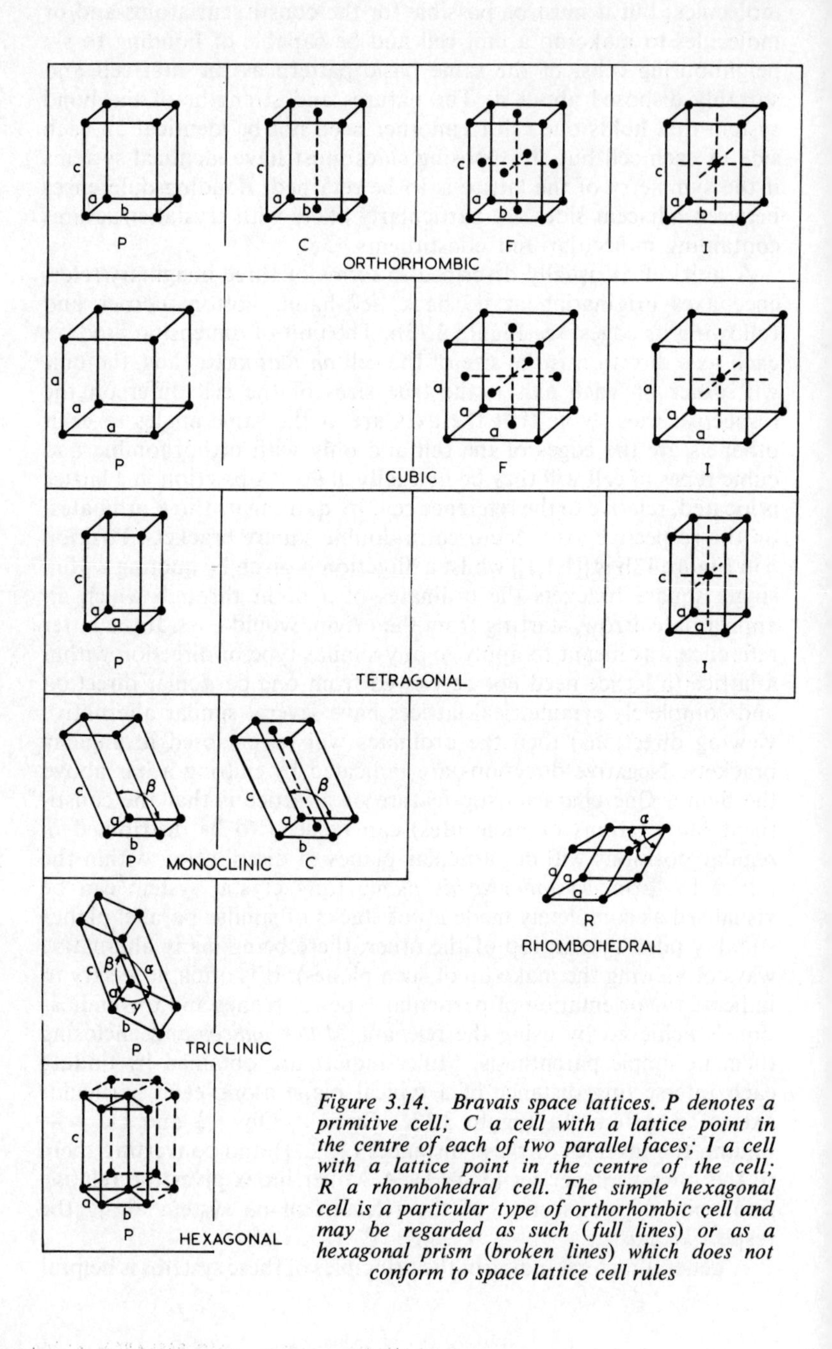

Figure 3.14. Bravais space lattices. P denotes a primitive cell; C a cell with a lattice point in the centre of each of two parallel faces; I a cell with a lattice point in the centre of the cell; R a rhombohedral cell. The simple hexagonal cell is a particular type of orthorhombic cell and may be regarded as such (full lines) or as a hexagonal prism (broken lines) which does not conform to space lattice cell rules

in studying the nature of crystals, but is not essential to the present level of discussion.

The fourteen possible types of space lattices, also known as *Bravais lattices*, are indicated in Figure 3.14. The system divides itself into two types of unit cell. In the first group are the seven *simple* cells in which similar atoms are located only at the intersection of space lattice lines i.e. at the corners of the unit cells. The seven remaining, more complex cells have atoms and/or molecules located (or centred) also on appropriate symmetrical positions such as the centres of lattice spaces or the middles of opposite side faces. Six of the simple cells are also sometimes classed as *primitive cells*, the rhombohedral cell being classed separately.

Looked at from the aspect of unit-cell structure the sodium-chloride structure of Figure 3.8 can be seen to need the unit cell configuration shown in Figure 3.15, with chlorine atoms at the cube corners and in the centres of the faces, a sodium atom at its centre and sodium atoms at the centres of each edge.

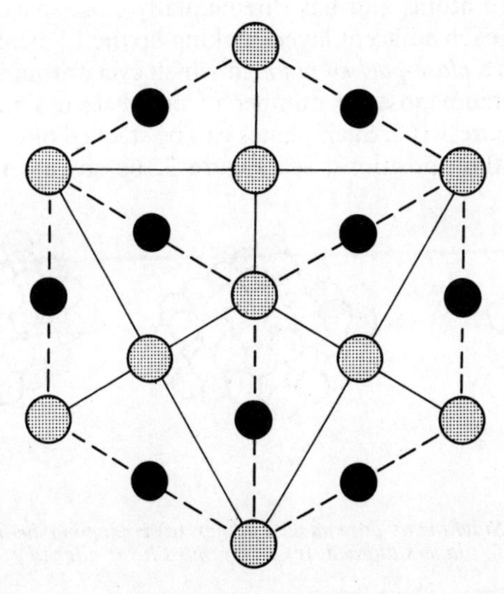

Figure 3.15. Sodium chloride unit cell (Na dark spheres)

One may look at cells from another point of view in relation to the number of *equal closest neighbours* possible for each atom position with each configuration. Thus in Figure 3.15 the number is six, four in one plane and two normal to it. When suitable atoms are trying to form into a crystal and the forces between the atoms are not specifically directional (as in most metallic bonding) the atoms will try to group themselves with the maximum density of packing relative to their atomic diameter and other characteristics, so the concept of closest packing becomes particularly important.

The face-centred cubic array (Figure 3.14) permits the closest symmetrical packing with 12 equal closest neighbours and the body-centred cubic array permits the next densest with 8 equal closest neighbours. The positioning of the latter is easily seen from the diagram, the centre atom having four neighbours above and four below, but the position is not so clear with the former. However if a corner atom is removed from a multi-cell face-centred cubic lattice structure three equal closest neighbours are exposed lying in an octahedral plane angle at 45° to the base and at right angles to the cube diagonal which runs in from the centre of the removed atom. If the three atoms are removed, a similar array containing more atoms will be found and so on if each successive layer is removed. Each layer within the crystal is identical to the others and every atom in each such layer is immediately surrounded by six similar closest-spaced atoms and has three equally close spaced neighbouring atoms in each adjacent layer, making up the 12 mentioned above. Each layer is a *close-packed plane* in which every atom is surrounded by the maximum possible number of equal-sized similar atoms as shown in Figure 3.16a. Such planes can be stacked one on top of the other in nestling positions, see Figure 3.16b so that the respective

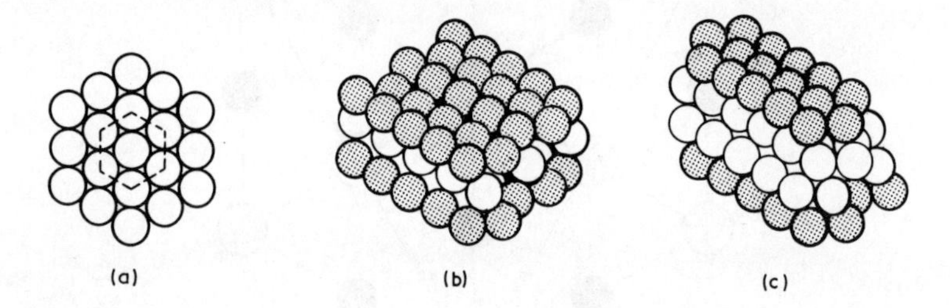

Figure 3.16. Stacking of close packed planes. (a) Hexagonal array. (b) Alternate layers aligned. (c) Every third layer aligned

atoms in all alternate planes are aligned with each other. According to the diagonal axis direction from which the structure is viewed eight similar systems of stacking orientation exist in a face-centred cubic lattice hence the name *octohedral* planes.

A similar closest-spacing but less symmetrical system of atom packing develops if close-packed planes are stacked in alternative nestling positions so that every *third* layer is in vertical alignment.

This gives a *close-packed-hexagonal* or *hexagonal-close-packed* array, see Figure 3.16c, which has only one system of stacking and is *not* a unit cell system since the middle atoms are not symmetrical with the others. The system can make up a hexagonal lattice if the atoms are looked at as paired diagonally with the centre of each pair on a corner of a hexagonal cell.

It is primarily because of the drive to achieve maximum density of packing in materials bonded mainly by metallic bonds that metals tend to crystallise into either the face-centred cubic, the close-packed hexagonal or the body-centred cubic form.

Where there is positive directionality of bonding between constituent units in crystals then other arrays are formed depending on the direction and nature of the forces, and on the relative diameters of the atoms if types are mixed.

3.18 MACROSTRUCTURE

The macrostructure of a material may be made up in a variety of ways. It may be made up entirely of one kind of molecule with each molecule randomly linked to its neighbours, or the molecules may be clustered into individually-ordered groups which themselves are randomly linked to each other. There may be more than one kind of molecule, with each molecule randomly linked, or there may be more or less complex clustering, with random linking of clusters. Alternatively, numbers of crystals may form randomly and be more or less firmly linked to each other across their interface or different types of crystals may be randomly or preferentially arrayed in conjunction with each other. In still other circumstances crystals, molecules and perhaps isolated atoms may be linked into a heterogeneous assembly.

Mixed structures of these various kinds usually give properties corresponding to an averaging interaction of the various factors. Thus a reactive material may be made less reactive by combining it with a less reactive material and so on. However, many properties are affected by more than just the simple juxtapositioning or sizing of molecules or crystals and the nature of the bonds between groups of constituents may have a major effect on properties. Thus macrostructural properties will be affected by

(1) individual properties of constituent groups and units
(2) size, shape and distribution of individual groups and units
(3) orientations of groups and units relative to each other
(4) effectiveness and nature of bonding between groups

Roughly speaking there are three types of bonding of constituent groups, aggregation, agglomeration and intercrystalline bonding and each of these is considered below.

3.19 AGGREGATION

Many molecules or coherent groups of particles will not bond to each other by more than very weak van der Waals' forces, therefore, if such substances are to become constructional materials they must be bonded into a non-separating liquid or into a stable solid, whichever suits the intended application.

Liquids need not concern us at this stage, but the solution of that problem would be to find a suitable carrier fluid, which should either itself bring some contribution to the desired external properties, or should be neutral but sufficiently viscous, or internally reactive, to hold the constituents in suspension in the desired relationship with each other.

In the case of solids some form of cement is used to stick the particles together to give what is commonly called an *aggregate* (although other names are sometimes used). The bonding agent must do several things

(1) hold the constituent particles in their required geometric relationship
(2) give sufficient strength to make the resultant solid usable
(3) either be neutral relative to the external chemical properties required from the material or contribute something useful to them.

Points (1) and (2) may be achieved in one or both of two different ways.

The bonding agent may react with and bond strongly to the surfaces of the constituent groups or particles so that it forms a bridge bond between them. In doing this it must not react so deeply into the

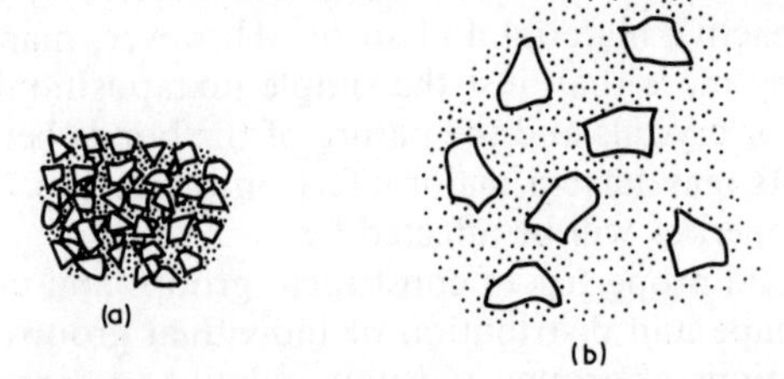

Figure 3.17. Aggregation of heterogeneous structures. The matrix may form (a) a thin strongly bonding layer between particles or (b) a self supporting background in which the particles are embedded perhaps with very little bonding

particles that it destroys their nature and it must be itself strong enough to give the essential cohesive strength. Generally, very little relative bulk of such a cement is needed to give effective aggregation (Figure 3.17a) since the structure is integrated mainly by interbonding.

Alternatively, the agent may form a self-supporting network into which the basic particles are embedded with little significant bonding. The agent must not introduce harmful chemical properties to the aggregate and must be strong enough to give a large proportion of the requisite mechanical strength. A cement of this type is required in relatively large volume (Figure 3.17b) compared with the first type.

Builder's mortar and concrete are common examples of aggregations, the basic particles being sand and sand and gravel. Abrasive wheels are another example.

3.20 AGGLOMERATION

Many forms of basic constituents are able to bond more or less strongly to each other if their surfaces are suitably prepared and treated in contact with each other. Sometimes the process, called *agglomeration*, is facilitated by the presence of a foreign constituent which reacts with the surfaces and prepares them for bonding. Frequently heat and pressure are used as activating agencies and melting of the surfaces may provide a mechanism by which mutual fusion can occur. Alternatively, the mechanism may be a process of local interdiffusion between surface atoms aided by suitable particle distribution so that the maximum surface area contact is achieved with the minimum of entrapped gas from the atmosphere. Another method is by deliberate surface oxidizing or other chemical reaction to induce a surface condition suitable for bonding between the products of the reaction. Probably the commonest applications of this method are found in the firing of pottery, china clay and other ceramics.

3.21 INTERCRYSTALLINE BONDING

Individual crystals are often difficult to bond effectively to each other, but other kinds bond readily. Most metals for example, solidify from the molten state into a *polycrystalline agglomerate* made up of numbers of individual crystals randomly orientated but firmly bonded to each other across their interfaces. In some cases the interfacial bonds are differently distributed compared to the bonds within the crystals but in others it is difficult to detect any marked

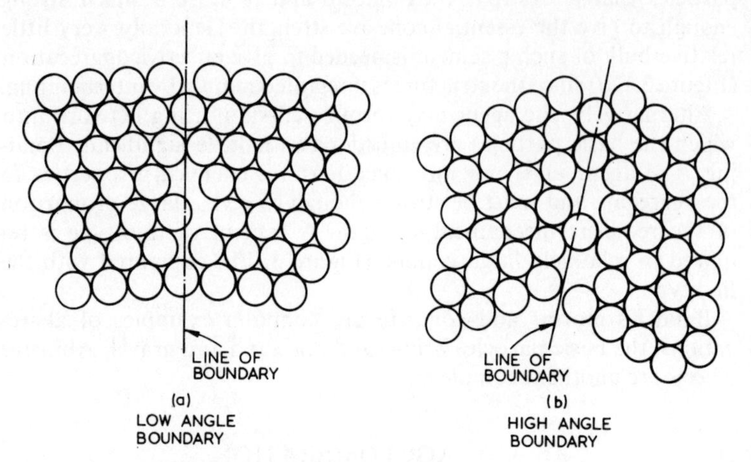

Figure 3.18. Bonding faces between crystalline grains. Poorer fits with greater angles between orientations

differences. In fact the bonding mechanism is usually of the same type but the bonds are less closely and regularly spaced, because the atoms or molecules are not ideally orientated. (Figure 3.18) and the higher the angle of misalignment the less easy the matching. When forming intercrystalline bonds the interfacial atoms are likely to be displaced significantly from their normal stable positions and it may be impossible to say to which side of the interface a displaced atom truly belongs.

The latter fact has a strong influence on the behaviour of crystals of this kind in the presence of heat since it may be possible for the thermal excitation to encourage interfacial atoms to attach themselves more strongly into the lattice on one side or the other. This causes the interface to move and one crystal to grow at the expense of the other, giving a phenomenon known as *grain growth*. The less distorted crystal face is the one most likely to grow. In some cases crystals orientated in a specific direction relative to the mass are the ones to grow preferentially and, if the process is carried to its logical conclusion, can lead to the structure reforming into a smaller number of larger crystals all orientated more or less in one similar direction, a state known as *preferred orientation*.

Crystals with relatively mobile surface atoms (most metals), in contact across an interface are amenable to joining by mutual diffusion of surface atoms under heat, either by simple melting diffusion or by heat-assisted solid-phase diffusion, a facility widely used in welding.

Intercrystalline faces are usually the most weakly bonded parts of a polycrystalline structure. Foreign atoms or molecules can greatly affect their nature in some cases strengthening them, but more commonly weakening them particularly if the foreign atom or molecule is insoluble in (i.e. will not join into) the parent lattices.

Size of individual grains influences all properties but particularly strength, a small average grain size giving notably greater strength under normal conditions than a large average grain size.

BIBLIOGRAPHY

MARTIN, J. W., *Elementary Science of Metals*, Chap. 3, Wykeham, London

MOFFAT, W. G., PEARSALL, G. W., and WULFF, J., *Structure and Properties of Materials*, Vol. 1, Chap. 1, Wiley

TWEEDDALE, J. G., *Metallurgical Principles for Engineers*, Chap. 3, Iliffe, London (1962)

VAN VLACK, L. H., *Elements of Materials Science*, Chaps 2 and 3, Addison-Wesley (1968)

4 Basic and Significant Properties

Of the basic physical properties, the most obviously important both to manufacturer and user are those called *mechanical properties* which relate the reactions of material to the application of external forces.

Chemical properties are just as important as physical properties but the properties themselves are not open to simple definitions and their significance is often difficult to discern.

4.1 MECHANICAL PROPERTIES

The *mechanical properties* of a material are the relative resistances of the material to the application of basic types of mechanical force. It should be understood that to look at mechanical properties in this way can be artificial because in service, the mechanical forces are rarely of a simple nature. Different types of force are usually acting simultaneously and are often widely and rapidly varying in intensity and direction. However, bearing this in mind there are three primary types of forces (see Figure 4.1)

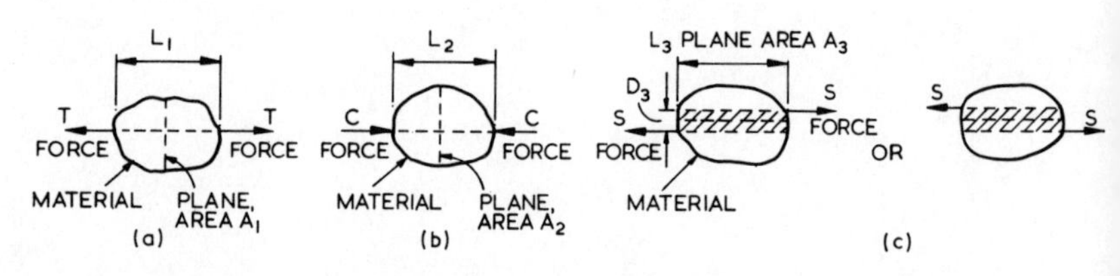

Figure 4.1. Basic force modes (a) tension (b) compression (c) shearing

(1) *Tension*—to pull the structural fabric of a material apart
(2) *Compression*—to push the structural fabric of a material closer together
(3) *Shearing*—to cause one part of the material to slide over another

Despite the artificiality of isolating these basic forms, this is still the only way in which simple results can be derived.

To use these basic forms of loading three approximately true assumptions are made

(a) It is assumed that force is transferred in a simple manner from the load source to the material resisting the force
(b) It is assumed that the material will resist force in a simple uniform manner. (This must be untrue since no material is perfectly homogeneous and isotropic)
(c) It is assumed that all material will always behave in the same manner

Using these assumptions and referring to Figure 4.1. The tensile force T will act uniformly within the volume between the points of external application and the internal reaction to it will be a *direct stress* σ or f i.e. an average reaction force per unit transverse area equal to the load force T divided by the relevant transverse area A_1 giving

$$\text{tensile stress } \sigma =_t \frac{T}{A_1}$$

in suitable units such as Newtons per square metre (N/m²). Simultaneously, the dimensions of the material will change in a uniformly proportionate manner. Such a proportionate change of dimensions is called *strain* (ε or e) and is measured as proportionate change of dimension related either to the original load-free dimension (*engineer's strain*) or to the dimension preceding the last increment of load (true strain). Thus in the former case the original L will increase say by l_1 giving the ratio

$$\text{axial tensile strain } \varepsilon_1 = \frac{l_1}{L_1}$$

Associated with this tensile strain will be an appropriate transverse contraction of the area A_1 for which a proportionate diametral change of dimension (*transverse strain*) negative this time, may be found in the same way

$$\text{transverse contractional strain } \varepsilon_2 = \frac{d_1}{D_1}$$

where d_1 is the reduction in the transverse dimension and D_1 is the initial transverse dimension.

Depending on the elastic and/or plastic properties of the material the transverse strain will have an approximately constant ratio relative to the corresponding axial strain in the same segment of material. This ratio

$$\frac{\varepsilon_2 \text{ transverse strain}}{\varepsilon_1 \text{ axial strain}}$$

known as *Poisson's ratio* (ν) depends on the relative proportionate change in the volume of the material induced by the application of the load. Cross-sectional shape influences transverse strain distribution, therefore Poisson's ratio applies only to uniformly circular cross-sectional areas.

Similar compressive direct stress and strain values may be derived.

$$\text{compressive stress } \sigma_c = \frac{C}{A_2}$$

and axial compressive strain (negative)

$$\varepsilon_c = \frac{\text{axial contraction } l_2}{\text{original axial length } L_2}$$

The corresponding Poisson's ratio will be similar.

The situation with shearing is different to that for tension and compression. Stress and strain are developed only in the shaded volume between the planes of action of the force components.

$$\text{Shear stress } \tau \text{ or } q = \frac{\text{shearing force } S}{\text{area of application } A_3}$$

$$\text{Shear strain } \gamma \text{ or } \phi = \frac{l_3}{D_3}$$

where l_3 is the change in the axial distance L_3 between the external

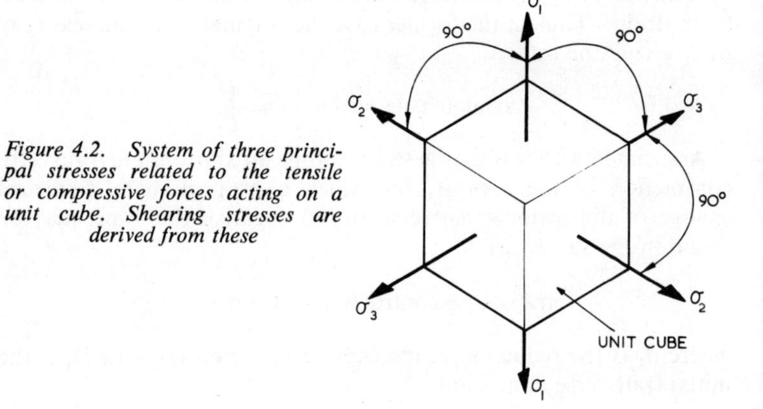

Figure 4.2. System of three principal stresses related to the tensile or compressive forces acting on a unit cube. Shearing stresses are derived from these

points of application and reaction of the force components and D_3 is the original perpendicular distance between their planes of application.

It is necessary to appreciate four facts concerning these basic modes.

(1) Any condition of loading can be broken down and its effect on a given segment of material analysed in terms of a unique simple combination of the two kinds of direct stress in a system of *principal stresses* (σ_1, σ_2 and σ_3) acting at right angles to each other on a unit cube of the material as shown in Figure 4.2. The resultant shearing stresses are found from the interactions between the principal direct stresses.

(2) Basic stresses such as those outlined are never developed in isolation. Tensile or compressive stresses have resultant shear stress components as shown in Figure 4.3a, the intensity of the shear stress component being a maximum at half the intensity of the principal stress. Shear stresses have resultant tensile and compressive components as shown in Figure 4.3b, the intensity of tensile and compressive components each being equal to that of the shear stress.

(3) Differing modes and conditions of application of basic load systems can produce different effects in the same material and these effects can be very complex and difficult to study even in terms of principal stresses.

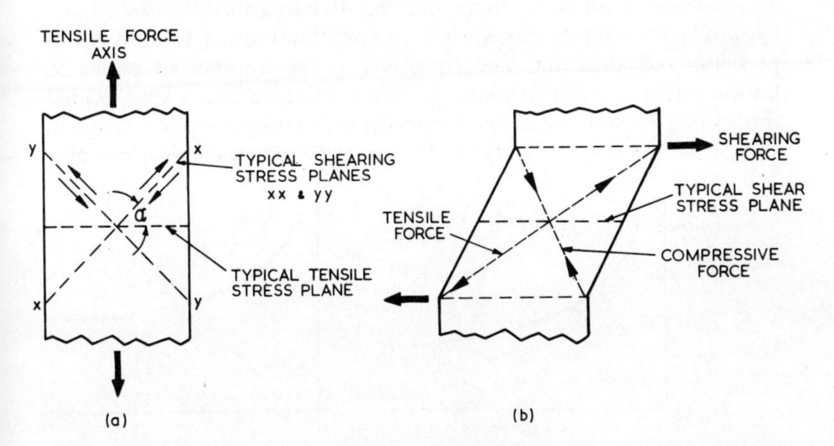

Figure 4.3. Resultant force and stress components (ft = tensile stress, fc compressive stress and fs shearing stress). (a) Shearing components of tensile forces. Every direct stress has shearing components. Maximum shear component occurs at α = 45, and is ft/2. (b) Tensile and compressive components of shearing forces. Maximum occurs in planes transverse to shearing direction and is fs = fc = ft

(4) Conditions of compression and shearing are difficult to study rigorously, therefore precise property values are not readily assessed for them.

From these facts the current situation in the practical study of materials has arisen. Only tensile properties are easy (and hence cheap) to assess readily and thus tend to form the basis of design and application. This situation is accentuated by the fact that failure in tension is likely to be the most critical kind of failure in service. Compression failures and shearing failures are not normally likely to give such abruptly catastrophic failures. Consequently there is a strong emphasis on tensile properties when studying mechanical properties and the engineer tends to allow for his ignorance of these properties by a reasoned increase in the nominal reserve of strength (*safety factor*) he tries to build into his construction. Because the effects of fluctuating and special conditions of loading are difficult to assess, these effects have to be estimated by empirical systems of comparative approximation. As a result, mechanical properties include such concepts as *fatigue strength*, *creep strength*, *notch impact strength*, *crack opening displacement* and *hardness*. Of these, the first four may be considered in relation to the normal basic forms of loading but the last is a somewhat vague and ill-defined property although of great usefulness for many practical purposes.

FATIGUE STRENGTH is the resistance of a material to frequent application or fluctuation of a load cycle. It may be specified as an *endurance* an *endurance limit* or a *fatigue limit* relative to a *specific kind of load cycle* which could be any variety of combinations of tension, compression and shearing. The *endurance* is the number of cycles to failure under a *specified* stress cycle at a *specified stress* level, whilst the *endurance limit* is the maximum stress intensity, or stress range in some cases, of a specified type of stress cycle which will just not give

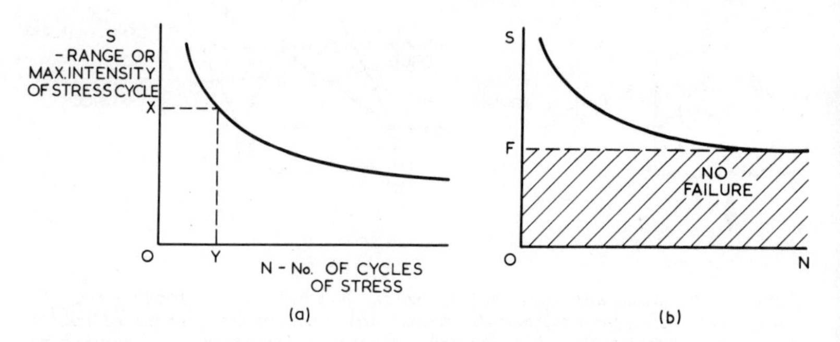

Figure 4.4. Graphical recording of fatigue properties. (a) Typical S/N curve OX = endurance limit for Y cycles. OY = endurance for OX stress. (b) Fatigue limit. No failures below F

failure in a *specified* number of cycles, see Figure 4.4a. The former is meaningless without knowledge of the type of cycle and relevant stress levels and the latter is meaningless without knowledge of the type of cycle and number of cycles. The *fatigue limit* has a specific meaning confined to certain materials which, even when subjected to an infinite number of a particular type of cycle of stress at stresses lower than a maximum called the *fatigue limit*, will never give failure (Figure 4.4b). The importance of fatigue properties in any application where fluctuating stress is applied, or where vibration is present, will be realised when it is appreciated that repeated applications of a load appear to reduce the strength of a material. In general, with normal metallic load-bearing materials neither the speed of application of load nor the frequency of cycling within the particular range seem to be of critical importance to strength, but these aspects must be considered in relation to the use of less established and newer materials. Ten million cycles is often taken as representing a normal lifetime of service but this assumption should be treated with caution.

Fatigue strength is invariably lowered in the presence of a stress-raising discontinuity such as a notch or a sharp change of shape of a member.

CREEP STRENGTH is the resistance of a material to conditions (usually elevated temperature) in which a steady load is likely to induce continually increasing plastic (irreversible) strain with time of application. It may be defined (1) in terms of the stress required to give the maximum tolerable *rate* of increase of strain relative to a specified temperature after a given time (2) in terms of the stress required to give a maximum tolerable *total amount* of strain after a specified time at a specified temperature and (3) in terms of the maximum stress that will just fail to give *failure* in a specified time at a specified temperature. Each of these criteria can be adapted to represent critical conditions for particular applications. Therefore an infinite variety of limits is possible depending on requirements. Commonly the time scale is limited to 10^3, 10^4 or 10^5 hours representing short, medium and long-time service. Assessment is costly and time-consuming and results have to be summarised in forms similar to that shown in Figure 4.5.

Thermal cycling is a factor which may have to be taken into account with creep in conjunction with the possibility of *thermally induced* fatigue failure due to differential expansion and contraction.

NOTCH IMPACT STRENGTH can be very important when material of irregular shape or particular sensitivity to cracking is likely to be subject to shock loading. Many materials are very sensitive to localised stress concentrations induced by particular loading conditions or by disturbances in continuity of shape, such as notches, and

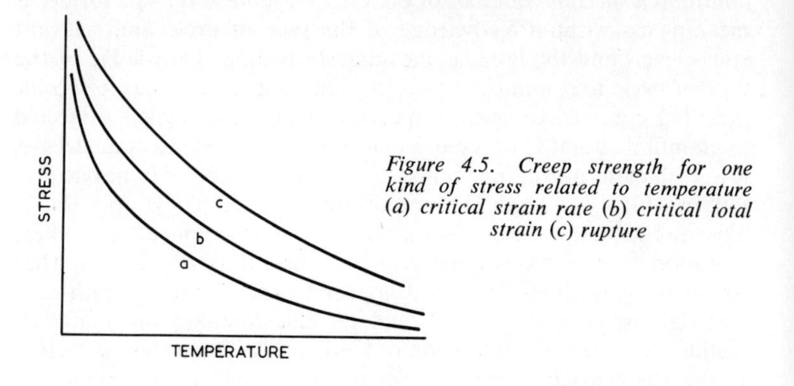

Figure 4.5. Creep strength for one kind of stress related to temperature (a) critical strain rate (b) critical total strain (c) rupture

are likely to give abrupt catastrophic failures if subjected to suddenly applied loads. Most derived values for such properties are simply empirically comparative and are usually given in terms of the energy (Joules or kgfm) required to fracture a standard test piece with a standard shape of notch, at a fixed rate of deformation but at differing temperatures. Results are given on a graph of the type indicated in Figure 4.6 which shows two sets of results, one for a material

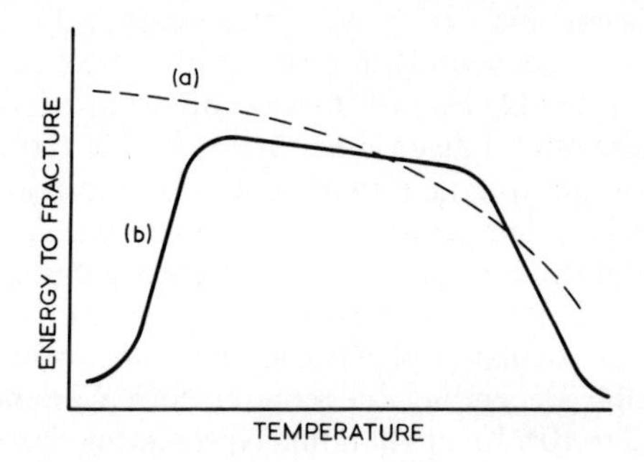

Figure 4.6. Notch impact tests (a) is a material insensitive to temperature (b) sensitive to high and low temperature

sensitive to temperature change and one for an otherwise similar material insensitive to temperature change. Results are most commonly quoted in terms of a test called the Charpy V-Notch Impact Test, but other test systems are also used.

CRACK OPENING DISPLACEMENT is closely related to notch impact strength and is another way for assessing sensitivity to stress and strain concentrations. The *COD* is the proportionate amount of increased opening of a controlled form of notch at the stage at which a crack begins to open up at the notch root under slow loading at a given temperature. Results are given as direct loads required to

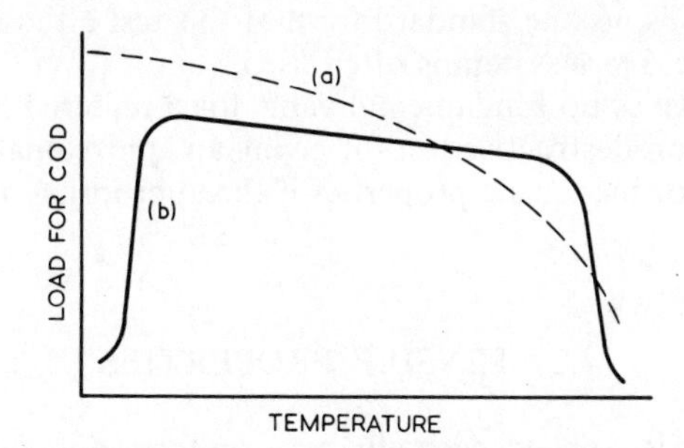

Figure 4.7. Recording of load for initiation stage of fracture (COD) as shown by strain at a standard notch under steady load (a) insensitive material (b) hot and cold sensitive material

give critical opening in a similar manner to notch impact strength, as shown in Figure 4.7 but the lower and higher transition temperature levels are likely to be lower and higher respectively.

HARDNESS is difficult to specify since its meaning depends both on the material being tested and on the method of test. For stiff plastic materials it is usually derived from some form of indentation test made by pushing a standard penetrator into a suitable surface of the material under a specific force for a specified minimum time and then deriving the surface area of the resulting indentation and dividing that into the load. The most standard form of test is the Vickers hardness system performed by pushing a standard square pyramid diamond indenter, apex first into the surface of the material, calculating the relevant value of surface area of the indentation that is produced and dividing that into the load. Results are specified as say 200HV30 where HV is the identification of the test system (hardness-Vickers) 30 is the force in kgf put on the indenter and 200 is the value obtained by dividing 30 kgf by the surface area of the indentation. There are other varieties of such hardness tests using a ball indenter (Brinell system, symbol HB) or different specified indenters (Rockwell system using a ball or a diamond cone indenter, symbol HR with a letter added to identify the condition of load and indenter shape). Brinell values are approximately equal to Vickers values but suffer from the effects of variation of indentation shape with differing depths of indentation. Rockwell values are arbitrary directly-read values taken from the depth of the indentation.

Highly elastic or very brittle materials are not amenable to such indentation testing, but can be tested with some consistency by allowing a shaped hammer weight to fall from a specified height on to a horizontal surface of the material and then measuring the hammer's height of rebound to give a measure of the *elastic hardness*. At present there is no one standard form of this test although a system known as the *Shore* system is often used.

Hardness gives no fundamental value for a material but can be a very useful non-destructive test for giving an approximate indication of the value of the tensile properties if the appropriate relationships are known.

4.2 TENSILE PROPERTIES

If a tension is applied coaxially to a uniform and symmetrically shaped piece of material with some elasticity, such as the solid cylindrical test piece illustrated in Figure 4.8, then the values of stress and strain at various stages of increasing or decreasing load can be measured, given the necessary equipment.

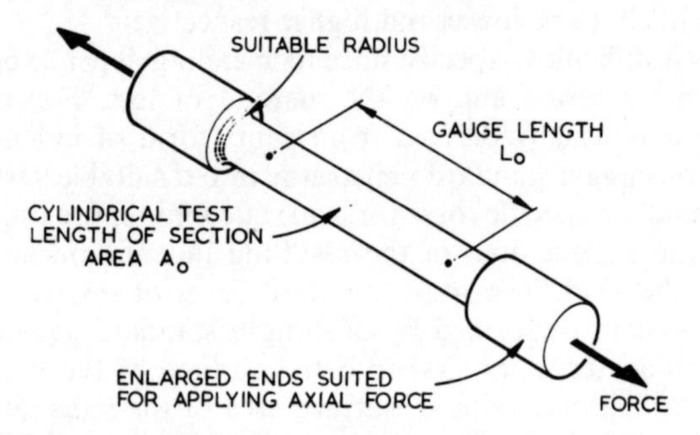

Figure 4.8. A tensile test piece

Engineering stress (load/original area of cross-section) requires only that the initial specimen size be known and that the 'load' be measured at each stage of the test so that the stresses can be calculated. For either true stress or transverse strain to be calculated requires careful measurement to be made of the gauge diameter, at each stage of a test to allow for change, possibly irregular, in the area and shape of cross-section of the specimen. In this text, except where specified, engineering stress is always implied.

Axial strain is relatively easy to measure, requiring only an instrument for measurement of change of axial length during a test. From the derived values it is possible to calculate the engineering strain and also the true strain, but usually engineering strain, is the one required and this is implied in the text except where shown differently.

Having derived the respective stress and strain values it is possible to study the tensile characteristics of a material from a graph comparing these values—a *stress-strain* diagram—and to specify tensile properties by specifying notable features of the relationship. All solid materials have some degree of *elasticity*, that is the material will tend to return towards its original dimensions as the deforming load is reduced or removed. Many materials are *plastic* and tend to retain some of their change of shape permanently after a deforming load is removed. Plasticity under tension appears as a *ductile* deformation—a permanent increase in length accompanied by a reduction in transverse area. Elasticity merges into, and its effect may be lost in any significant amount of plastic deformation which develops. All materials will break if stretched far enough.

The nature of each of these phenomena will tend to be characteristic for a material and may be specified more or less rigorously from an appropriate stress-strain diagram. One typical form of diagram often encountered in metallic materials is shown in Figure 4.9. If a material with these characteristics is loaded only into the range *Oa*

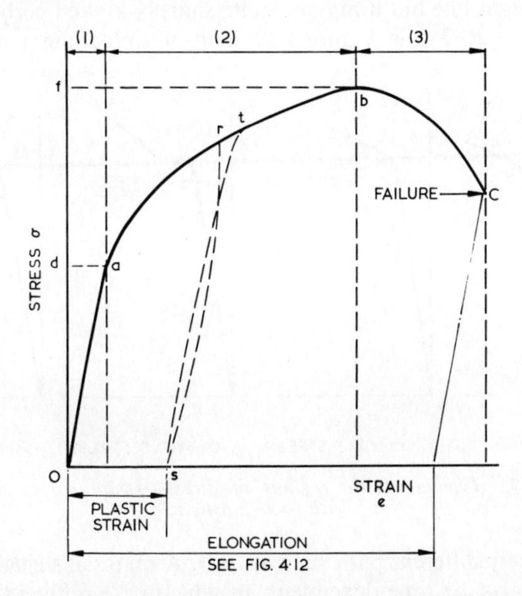

Figure 4.9. Stress—strain diagram for a ductile material (1) elastic deformation (2) elastic + plastic deformation (3) elastic + plastic + creep to failure. Od = elastic limit, Of = Maximum stress = tensile strength

the material will recover its original shape along line *aO* on removal of the load, but if it is loaded into the range *ab* then unloaded, from say point *r*, the diagram will have to be redrawn as shown from *r* to *s*, *Os* representing the proportionate permanent change of dimension or *plastic strain*. *s* would become the new zero and reloading would follow line *st* eventually joining with the original graph at *t*. If loading is taken into the range *bc* and a steady load left on, failure would result because the material would creep until failure occurred. Note that the latter stage does not give a true picture of the real behaviour of the material but, with skilled interpretation it can give some indication of the likely type of failure.

Should the elastic behaviour be of the proportionate type suggested in the diagram for range *Oa*, then elastic behaviour may be readily specified by the tensile *elastic modulus* or *elastic rate E* perhaps better known as *Young's Modulus of Elasticity*, which is the rate of slope of the straight line *Oa*, the theoretical stress required to induce 100% elastic strain (100% elastic strain isn't possible in most cases). However elastic behaviour is not always as simple as this. With some materials including some metals stage *Oa* would be a curve and not a straight line, but unloading would follow back more or less on the same path to *O*. With other materials not only is *Oa* not a straight line but it may be quite sharply kinked perhaps on the lines shown at *Oxa* in Figure 4.10a and on unloading it may follow

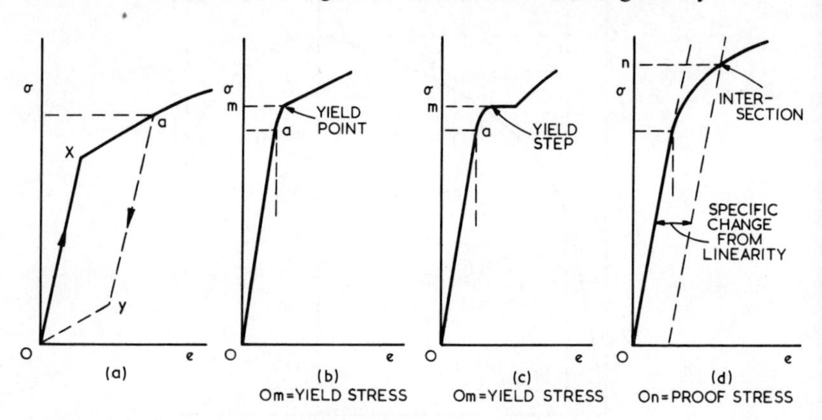

Figure 4.10. Typical elastic regions of stress—strain curves of materials with some plasticity

a completely different path such as *ayO*. A material's elastic properties may also be time-dependent, in which case different speeds of loading and unloading could produce different curves. Many non crystalline materials are of this latter kind making it difficult to specify their elastic properties and difficult to incorporate them into structures. It is self-evident that a material with a low modulus deforms to a greater extent under elastic loading to a given stress, than does a material with a high modulus under the same stress and this difference in behaviour could be critical for many practical applications.

Associated with the completely elastic part of a stress-strain diagram is also the *elastic limit*, the point *a* in Figure 4.9 represented by the stress *Od* which is the maximum stress at which no significant plastic deformation occurs. This point is often very difficult to determine accurately, so a compromise is accepted by the engineer and the stress on the curve at which either a sudden marked increase in *plastic* strain called a *yield-point* appears (see Figure 4.10b and c) or the stress at which a specified departure from linearity of the curve occurs (see Figure 4.10) is used. The equivalent stress for the former is called the *yield stress* and for the latter the *proof stress* which must be qualified by the amount of non linear strain it represents, e.g. 0·1% *proof stress* (0·1%, 0·2% and 0·5% strains respectively are probably the amounts of non linear strain most commonly used for this purpose). It should be noted that the yield stress is slightly above the elastic limit since it clearly represents the incidence of gross plastic strain, but a proof stress could be below the elastic limit if the elastic part of the diagram is markedly non linear. The criterion for proof stress is determined by the amount of non linear behaviour tolerable to the user.

With the onset of plastic deformation the features of stress-strain diagrams are less predictable and marked divergence between the conventional stress-strain diagram and a true stress-true strain diagram appears. Figure 4.11 shows the tendency; the normal

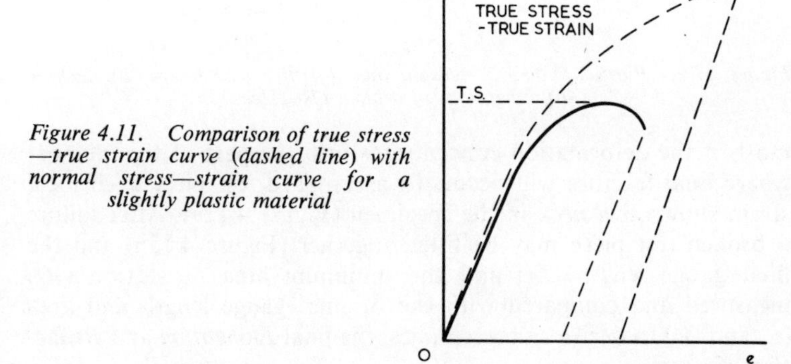

Figure 4.11. Comparison of true stress —true strain curve (dashed line) with normal stress—strain curve for a slightly plastic material

diagram (full line) if plastic flow occurs, will tend to show a maximum followed by a falling off, but a true stress-true strain curve (dashed line) will show a continuing rise in stress because the specimen thins down more rapidly than load rises. For the same reason true strain will rise above the engineering strain because the latter is an average general ratio whereas the former is associated with the maximum local strain at the point of failure, where the specimen thins most. The maximum reached in the normal diagram is a convenient value for the engineer to use as a measure of the strength of his material and is called the *tensile strength*—T.S. (it was formerly called the ultimate tensile strength—U.T.S. but the term is misleading and has been dropped).

To those workers concerned with deliberate deformation of material the slope of the true stress-true strain curve in the plastic range is a phenomenon of great importance, since it enables them to estimate the forces involved in producing specific changes in shape by controlled application of tension.

It will be noted that the third stage of Figure 4.9 does not appear in a true stress-strain diagram (see Figure 4.11) because its presence is due entirely to the combination of non uniform behaviour of the material under test with the engineering method of assessing average values for stress and strain. Typically, as final plastic deformation of the material begins, some parts of the member will be more intensely loaded than others, or will be more free to deform. Subsequently

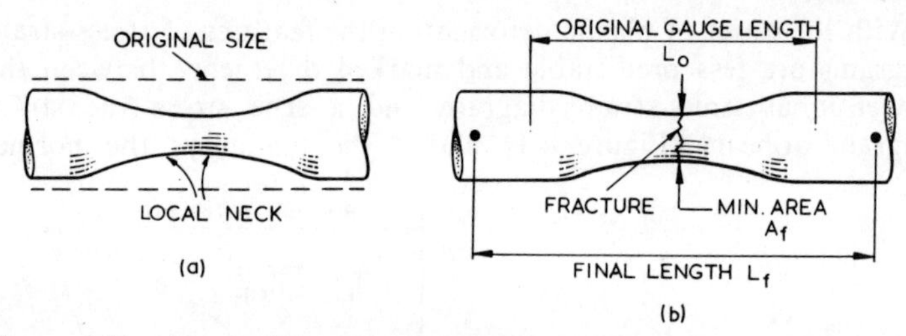

Figure 4.12. Plastic failure in tension and ductility assessment (a) necking (b) dimensions of broken test piece

most of the deformation concentrates into the part of the member where final fracture will occur. In a standard test piece such local strain shows as a *neck* in the specimen (Figure 4.12a). After failure a broken test piece may be fitted together (Figure 4.12b) and the final gauge length (L_f) and the minimum area of section (A_f) measured and compared with the original gauge length and area (L_o and A_o) to derive as percentages, the final *Elongation* and *Reduction of Area*

$$\text{Elongation} = \frac{L_f - L_o}{L_o} \times 100\%$$

$$\text{Reduction of Area} = \frac{A_o - A_f}{A_o} \times 100\%$$

These values in conjunction with the final mode of fracture can give a fair picture both of the *ductility* (plasticity under tension) and of the likely failure mode of the material. The greater the values of these ratios, the more the total plastic deformation. The greater the difference between them the more local is the necking that has occurred. (Reduction in area is usually greater than elongation.) Mode of fracture is too complex to consider at this stage but it should be noted again that tensile fracture is likely to be irrevocably and rapidly completed once it begins.

If a material is brittle, with no significant plasticity, it is very difficult to obtain realistic values for tensile strength. Fracture occurs whilst the material is still in the elastic range of stress, and experimental scatter is likely to be great because normally negligible variations in structure or shape can have great effect on strength. A compromise here is to use a simple bending test and work back from that to find the *Modulus of Rupture*, the breaking stress derived from *Bending moment at rupture* = *max. tensile stress* × *modulus of section*. (The modulus of section is the geometric relationship between the max. stress developed in a simple elastic material and the shape of cross-section of the test bar.)

Fatigue loading in tension is likely to be critically important in any construction that is subject to fluctuating loads or vibration, therefore the majority of specifications of fatigue properties include significant consideration of the effects of tensile components of force.

Creep in tension could be critical with respect both to deformation and failure, but particularly with respect to the latter since the creep strain rate in tension is always liable to increase suddenly towards fracture as the cross-sectional area of a creeping member progressively decreases.

The progressive nature of cracking under tension makes the likelihood of a total fracture from a geometric discontinuity, such as a notch or crack, greater under tension than with other types of stress. Therefore most notch sensitivity tests and all COD tests are performed with particular reference to loading in tension.

Consideration of mechanical properties, particularly tensile properties, is not complete without taking account of the influence of temperature. In general lowering temperature tends to raise tensile strength, yield strength and elastic modulus and to reduce ductility. Conversely raising the temperature tends to lower the former and increase the latter. Creep strength is invariably lower at elevated temperature. A limit is reached with respect to ductility as melting temperature is approached and melting begins. Structural change associated with temperature change in the solid state, may greatly modify the more general trends of temperature effects. Particular account should be taken of likely changes in notch sensitivity under tensile load relative to change in temperature of use of a material.

4.3 COMPRESSION PROPERTIES

It is very difficult to isolate the effects of compressive loads from other effects when studying materials. In fact with all except the most plastic of materials, the criterion of failure in compressive loading is likely to be tensile or shearing failure. If the specimen under test is relatively long ($L/D > 3$) failure will probably occur by buckling (Figure 4.13a). If the length is short ($L/D < 3$) then, because the

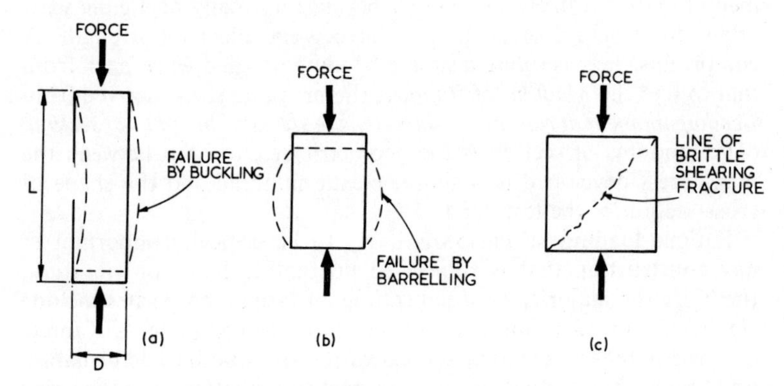

Figure 4.13. Failure modes in compression (a) instability (b) bulging under frictional restraint (c) brittle shearing

specimen end faces are frictionally restrained by the loading faces, a plastic specimen will *barrel* as shown in Figure 4.13b and this induces (1) hoop tension around the girth (2) axial side tension down the outside surfaces and (3) biaxial tensions at the centre as the outer diameter expands away from its centre. Failure in tension is possible in any one of these localities.

If the material under test is not plastic, or if its shear strength is low, and the L/D ratio is kept small then failure is likely by shearing along a plane direction such as that shown in Figure 4.13c since the associated shear stress component of a compressive stress is always half the intensity of the compressive stress.

If reasonable tests of compression strength are not readily made on materials with a certain amount of plasticity how can compression be allowed for in use? The answer is to use the equivalent tensile properties as a basis for calculation. The tensile strength, tensile yield strength (or its equivalent) and the tensile modulus of elasticity are used as the compression strength, compression yield strength and compression modulus of elasticity respectively. This is a safe practice since if there is likely to be any difference between the tensile and the true compressive properties the latter will always be at least a little greater in magnitude as shown in Figure 4.14. If the tensile

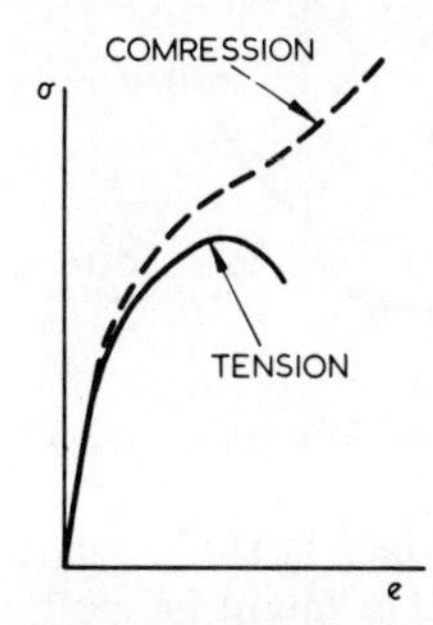

Figure 4.14. Relationship between tensile and compressive stress—strain diagrams for a reasonably plastic material. A very plastic material may collapse in compression without ever fracturing

and compressive strengths differ widely as they do in a very brittle material, property values may have to be derived by using direct compression tests in spite of their limitations.

Plasticity in compression is very important in relation to deliberate deformation of materials. Strains of 90% or more are possible in plastic materials under controlled compression. The compressive aspect of plasticity is called *malleability*—the capacity to deform by spreading under a compressive force.

Fatigue and creep effects in compression are likely to be similar to those in tension for reasonably plastic materials except that failure is much less likely. In brittle materials the difference may be extreme, with good compression fatigue and good compression creep properties and very poor complementary tensile properties. Change of temperature is likely to have effects on compressive properties similar to those on tensile properties, except that malleability will not decrease as rapidly as ductility as the melting temperature of a material is approached.

4.4 SHEARING PROPERTIES

It is almost as difficult to test for shearing properties as it is to test for compression properties, but for different reasons. In the first place it is very difficult to set up a simple uniform shear stress. In Figure 4.15a it seems easy to develop shear stress between the two knife edges in the plane *ab*, but too little material is involved for the accurate measurement of shearing strain. Even the small amount of

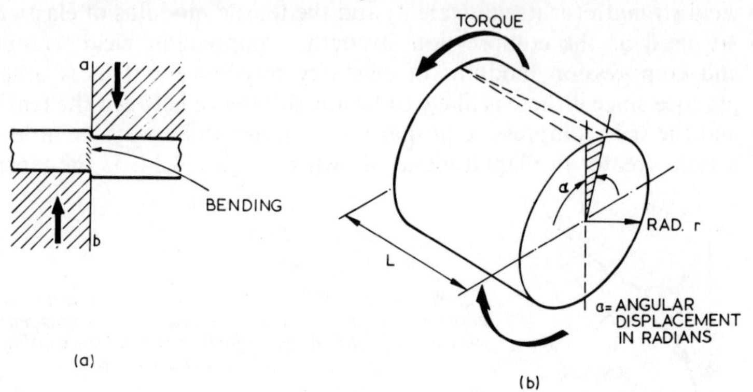

Figure 4.15. Testing for shear strength (a) simple shear (b) torsional shear

bending deflection possible between the knife edges is sufficient to upset any measurements of shearing conditions that might be made. Alternatively, it is possible to test a uniform cylinder of material in simple axially-centred torsion. In this case the principal stress component is torsional shear stress and a volume of material sufficient for measurement is easily arranged. Unfortunately the stress is not uniform and it varies from zero at the axis to a maximum at the surface. This situation may be eased by using a thin-walled hollow cylinder, but this then introduces problems of loading and of producing a cylinder suitably proportioned to avoid buckling. Fair results are obtainable with hollow cylindrical specimens, and it can be demonstrated that the pattern of stress-strain behaviour in shear is similar to that in tension.

Since the shear behaviour characteristics are similar to tensile behaviour characteristics it is to be expected that yields and strengths in shearing and tension respectively will be proportionate to each other for a given material and this relationship is used by the engineer, once the ratios are known, to derive from the tensile stress-strain diagram of a material the corresponding values of shearing yield stress, or proof stress, and shear strength. The *shear modulus of elasticity* (G) or modulus of *rigidity* is different in form from the tensile modulus of elasticity (E) but is related to it in the equation

$$E = 2G(1 + \rho)$$

(where ρ is Poisson's ratio) making G usually about one third of E. The shear modulus may be obtained from experimental values of torsional load and torsional strain measured on a cylindrical test segment such as that in Figure 4.15b. The strain (γ or ϕ) in this case is the appropriate circumferential surface displacement of one end of the test segment relative to the other end divided by the overall length (L) of the segment and the stress is derived from the applied torque (T) related to the size of cylindrical section by $T = \tau \times Zp$ where Zp is the *polar modulus of section* of the test piece relating the surface shear stress (τ) to the size and shape of the section.

There are always tensile and compressive stress components associated with a shear stress and if a material has limited plasticity, or low tensile strength, the tensile component may become the true criterion of failure and tend to give the steeply helical form of brittle fracture often encountered in torsion. A simple shearing fracture in torsion will tend to lie in a plane transverse to the axis of torsion. In a material with limited plasticity even a truly shearing type of failure will not be associated with much plastic strain but in a plastic material a shearing failure is likely to be associated with extensive plastic flow because it is shear stress that induces plastic flow and the shearing type of fracture in torsion does not open up as rapidly as a tensile fracture.

Fatigue properties in shear loading can be very important for many applications since critical fluctuating shearing loads often have to be transmitted through the members of a mechanism (e.g. the transmission shaft of a motor car). Account must be taken of these phenomena particularly to the ways in which the load can be applied. A torque may be positive or negative according to its direction of application (clockwise and anti-clockwise) and its significant cyclical combinations need study in relation to endurance, endurance limit and fatigue limit.

Creep is as liable to occur under shear loading as under tension or compression loading and the possibility of its occurring must be considered particularly if the temperature of service is elevated.

Materials subjected to shear loading, particularly shock loading, will be as sensitive to the presence of notches as are the same materials in tension, so this sensitivity factor should be considered.

Temperature change affects shear properties in a similar manner to its influence on tensile properties.

4.5 THE STRUCTURAL BASIS OF MECHANICAL PROPERTIES

Each mechanical property of a material is a measure, on a bulk scale, of the average reactions of the combinations of bonds that have formed within the material to the application of a specific type of external loading. In each case, the outcome is a result of interaction of

(1) types of bonds operative in each representative unit volume of material
(2) the relative strength and stiffness of each type of bond
(3) the numbers of each type of bond operating in each unit volume
(4) the relative groupings and orientations of the respective bonds in each unit volume

These interactions can be very complex and difficult to predict; nevertheless their general trends will follow the trend of the principal bonding mechanism that is holding together the constituent parts of each unit volume of material. Thus very powerful bonds may operate within each molecule of a material but, if the molecules of the material bond weakly to each other the mechanical properties of the material will be poor. Similarly, the internal molecular bonds of a material may not be very strong but, if the bonds between the molecules are numerous and reasonably strong, the material can be relatively strong.

Ideally every strong bond within a material should be able to make its full contribution to the strength of the material so that the overall bulk strengths are as near as possible in magnitude and nature to the sums of the strengths and natures of its individual bonds. This state is rarely attainable so there is usually a marked difference between the mechanical properties of the ideal and the real material, see Chapter 5.

4.6 ELECTRICAL PROPERTIES OF MATERIAL

When charged particles such as electrons or ionised atoms are subjected to an electrical field they tend to move in an appropriate direction within the applied field, see Figure 4.16. If a field is applied

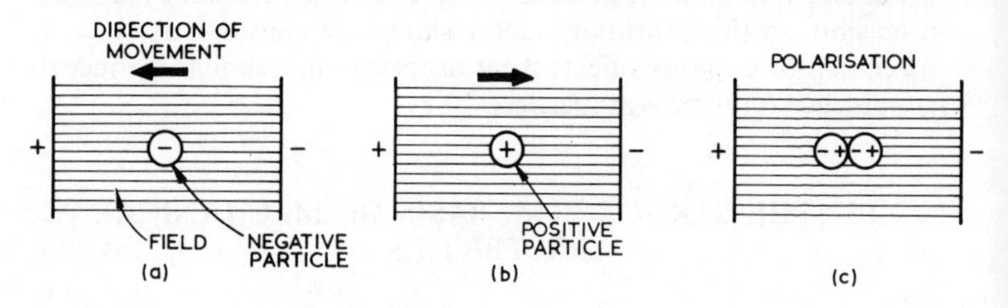

Figure 4.16. Effects of electric field on (a) negative particle (b) positive particle (c) stable atoms or molecules

to a material every charged particle in the material will be subject to the driving forces of the field and, if the field energy is enough to overcome the energy barriers to movement, charged particles may begin to jump from one locality to another in one direction within the material until the change in energy caused by the movement neutralises the field. Any general movement of charged particles in a material is called an electric current and any material in which significant current flow is possible is called a conductor.

There are always movements of charged particles within a material, even one with negligible conductivity, but these movements are contained by the energy barriers within orbital systems and between atoms as the various particles rebound against or react with each other. Hence the overall effect appears to be one of no movement. In the presence of an electrical field the movement of each charged particle is accelerated in one direction and retarded in the opposite direction. If the imparted energy is sufficient to overcome barriers to progressive movement there may be a drift or average movement of particles in one direction. Negatively charged particles will try to move in one direction and positively charged particles in the other. Thus, an electron (negatively charged) or a negatively charged atom tends to drive towards the positive pole in an electrical field. The resistance to movement of the respective particles differs widely between materials and even between different states of one material. Atoms find it difficult to move directionally in most situations and particularly in solids but valency orbital electrons are likely to be most free to move both because they are inherently more mobile and because they are likely to have the lowest energy barrier opposing them. Hence electron movement is the predominant effect in any electric current flow. However, even if electrons are mobile within a material they also interact against each other's energy fields and with the energy fields of the atoms in the material as they exchange orbits etc. Therefore although the imposition of an external electric field does cause accelerated movement in the direction of the field the rate of interference between particles also increases and there is a *resistance* to general uncontrolled flow of electrons.

These effects are assessed in terms of certain standard electrical units. The unit of electric *charge* or quantity of electricity is the *coulomb* which is the quantity of electricity transferred by one ampere (see below) flowing for one second. Electrical energy is measured in units of the *joule* (J) found by multiplying the amount of charge that flows by the *potential difference* measured in units of the *volt* (V).

Thus, electrical energy (Q) = charge (C) $\times$ potential difference (V)

The *ampere* (A) is that amount of current which, if it is flowing in each of two long thin parallel conductors placed 1 metre apart in a complete vacuum sets up an electromagnetic force between them of 2×10^{-7} newtons per metre length. Resistance (R) to flow of current is measured in units of the *ohm* (Ω) which is the resistance that would generate heat at the rate of one joule per second when a steady current of 1A is flowing through it. A *volt* is the potential difference across a resistance of 1Ω through which a current of 1A is flowing. Hence for any current flow

$$V = I\,(\text{current in A}) \times R\,(\text{resistance in }\Omega)$$

Electrical power is measured in units of the *watt* (W) which is 1J per second and therefore $1W = 1V \times 1A$ and hence for any current flow $W = IV$.

From these relationships the common equations of electrical power are derived and we have $W = I^2Rt$ where t is time in seconds.

Note that in physics the *electronvolt* (eV) is a measure of energy not potential difference, and is used where the joule is too large $1\ eV = 1{\cdot}6 \times 10^{-19}$ J.

Materials such as metals bonded by metallic bonds have interconnected valency orbital systems within which their shared electrons are always in movement. The greater the number of valency orbital paths and the more direct the connections between them the more readily can electrons be shunted along them under the influence of an applied electrical field. Conversely the more isolated the valency orbitals of a material the greater the energy needed to cause electron flow and the higher the resistivity. Therefore, predominantly ionically and covalently bonded systems within which the respective valency orbital groups are relatively self-contained, have a very high resistivity. The latter type of material is usually called a *dielectric* or an *insulator*. Between these two there is the *semi-conductor* with intermediate conductivity or directional conductivity.

Under the influence of an applied electrical field dielectric materials show a tendency towards polarisation of the charges within their constituent molecules (Figure 4.16c). This means the valency orbitals tend to centre more on the side of the molecule adjacent to the positive electrical pole and the ion charge tends to centre more towards the side adjacent to the negative pole. Although no interchange of charged particles takes place across the molecular boundary, and no current flows, the change in charge distribution can be used to improve the storage capacity of a capacitor (an instrument for storing an electric charge on separate, parallel, oppositely charged plates) by using a suitable *dielectric* material to fill the space between the charged plates. The possible electric charge in this situation is increased in proportion to the *dielectric constant* of the material. The dielectric constant is an arbitrarily derived value relating the effect on the capacity of charge of the presence of the material under consideration compared to the situation with a vacuum between the plates.

Certain conductive materials show a change to *superconductivity*, a state of zero resistivity if they are cooled to very low temperatures near absolute zero. This property is not yet clearly understood.

4.7 MAGNETIC PROPERTIES OF MATERIALS

Three forms of magnetic behaviour can be found in most materials *ferromagnetism*, *paramagnetism* or *diamagnetism*. In the first form the material shows a strong attraction to the pole of a magnet, in the second a very slight attraction to a magnet and in the third a very slight repulsion from a magnet. For most practical purposes materials coming in the second and third groups may be regarded as non magnetic.

Bulk material is not inherently magnetic. The tendency is for any localised magnetic fields within a material to appear randomly arranged and to be self-cancelling. This internal balancing of

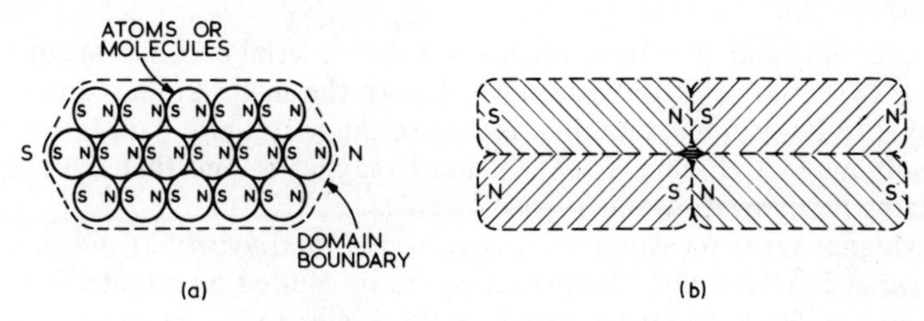

Figure 4.17. Formation (a) and neutralisation (b) of magnetic domains. Domain boundaries need not be grain boundaries

individual atomic magnetic fields cennot usually be effected by simple opposed pairing of the fields of adjacent atoms or molecules. Since atomic magnetic fields are directionally polarised and subject to normal opposite pole attraction and like pole repulsion, there is a strong tendency towards axial alignment of suitably situated adjacent fields, each alignment adding to the strength of the combined field, see Figure 4.17a, and tending to build up a *magnetic domain* within which the total magnetic field is polarised in one direction. Adjacent domains polarise in different directions oriented in such a way that the domains cancel out and the overall field is unpolarised. The size of a domain is governed by the balance between the natures of the magnetic fields of the atoms or molecules making up the structure of the particular material, but each domain is usually very small in relation to the bulk of material likely to be present in even very small components.

The magnetic orientations of the domains in some materials may be changed by external influences, notably external magnetic fields, see Figure 4.18 and in some cases by stress. Such changes may lead

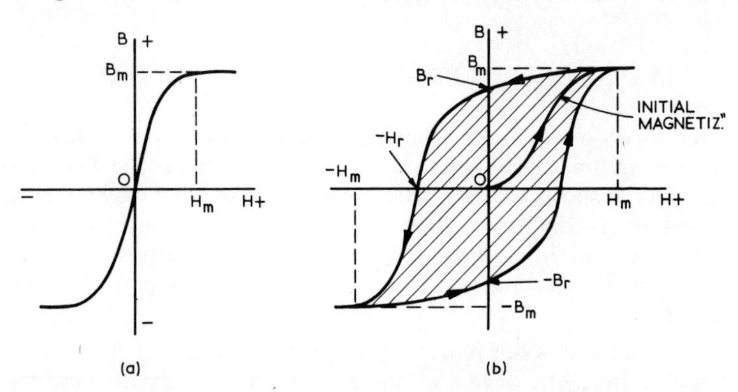

Figure 4.18. Effect of an applied magnetic field (H) on magnetisation (B) of different materials. (a) Soft magnet. (b) Hard magnet

to overall ordering of the domains and the material becomes magnetised. On removal of the external influence the material may quickly revert to its normal state, in which case the material is said to form a soft magnet (Figure 4.18a). A hard magnet is one that strongly retains its magnetisation (Figure 4.18b).

Magnetism is measured by assessing the *flux density* or *magnetic intensity* in terms of the force that would be exerted on a unit electric charge moving at unit velocity at right angles to the magnetic field at a specific point in the field in question. This value may be quoted in terms of *gauss* or unit lines of magnetic force per square cm; but in the SI system the *tesla* is used. A *tesla* is 1 Weber/m^2 and a *weber* (Wb) is a magnetic field intensity such that if a charge of 1 coulomb is moving at right angles to the field at a velocity of 1 m/s the charge is acted on by a force of 1N.

The flux density B in a material is related to the strength of a known field which is acting on it, by the equation

$$B = \mu H$$

where μ is the *permeability* or maximum magnetic-field-concentrating effect of the material and H is the strength of field that induces B. The permeability may be quoted as a *relative permeability* (μ_r) by quoting its ratio to the permeability of the same field (μ_o) in a vacuum. In the latter case the ratio μ_r is dimensionless. (Note that where no confusion is likely to be caused, the *relative* permeability is often indicated by the symbol μ.) For non magnetic materials μ_r is not likely to be high but for ferro-magnetic materials it can be 10^5 or more.

Associated with permeability is the *remanence*, the persistence of a magnetic field in a material after the inducing field has been reduced to zero. A hard magnet has a high remanence and a soft magnet a low value. Removal of such remanence B_r requires reversal of the inducing field ($-H_r$ Figure 4.18b) and progressive reversal beyond this point reverses the induced field to $-B_m$. Removal of the reversed inducing field and reapplication of the original field first causes a drop of remanence to $-B_r$, then to zero, and ultimately a rise back to B_m and subsequent similar cyclic changes then follow the same path through B_r, 0, $-B_m$, $-B_r$ back to B_m enclosing the so-called *magnetic hysteresis loop* the area of which represents the amount of energy lost during each cycle. When an electromagnetically induced field is changing rapidly, as it might do in a piece of electrical or electronic equipment, it is obvious that a very low remanence is desirable if energy loss and generation of heat is to be avoided. Conversely, a high remanence is desirable in a permanent magnet.

Strain may cause magnetisation or destroy it because, in some materials the magnetic field of each atom or molecule is associated with a dimensional assymmetry of the atom and molecule and a mobility of the orbital mainly influencing the magnetic field and the

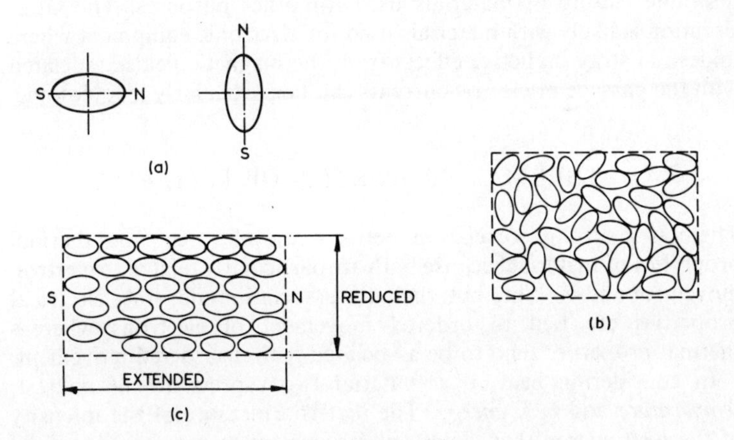

Figure 4.19. Possible dimensional effects of magnetisation (a) same atom with pole orientation switched 90° (b) material with random magnetic orientation (c) same volume magnetised

assymmetry. Usually the orbital size is elongated in the direction of the field therefore an applied tension may cause such orbitals to align with the tensile strain direction and hence to polarise magnetisation in that direction. A material of this kind will also tend to change its dimensions when it is magnetised in an external field, see Figure 4.19, but will usually revert to its normal size when the external exciting influence is removed. A hard magnet can be destroyed by stressing it in an unsympathetic direction. Increasing heat can destroy the magnetic properties of a material, the temperature at which the properties disappear being known as the *Curie* temperature.

Only a limited number of elements, notably iron, cobalt, nickel and gadolinium are inherently ferromagnetic. Some alloys of manganese and some ceramic compounds are also ferromagnetic but most other metals are paramagnetic and all non metals are diamagnetic.

Magnetic properties are associated with either the valency electrons in a material or with the electrons in an unfilled sub-shell. In the case of ferromagnetic materials it is parallel spin of a pair of electrons in an unfilled sub-shell caused by the local configuration of orbitals that gives rise to the strong effect, but in the other cases the relatively weaker magnetic field is a direct result of the weak magnetic fields resulting from the orbiting of the valency electrons, particularly if there is an unpaired valency electron.

Specific magnetic properties are frequently desired in a material; but in many other circumstances magnetic properties are an undesirable feature of materials used for other purposes. The latter situation is likely with materials used for electronic equipment where undesired stray inductive effects from the magnetic fields associated with the passage of electric currents can be particularly troublesome.

4.8 THERMAL PROPERTIES OF MATERIAL

There is a strong correlation between the electrical and thermal properties of material because both are particularly related to electron movement and mobility but, there the association ends since electrical properties are tied to ordered movement of electrons whereas thermal properties tend to be associated with disordered movement.

In considering heat in a material two aspects are of interest, *temperature* and *heat energy*. The first is a measure of the intensity of the heat energy that is present (equivalent to pressure in a fluid system). The second aspect is the actual quantity of heat energy involved (equivalent to volume in a fluid system).

The three concepts required to relate these aspects to a material are (1) *heat capacity* (2) *thermal conductivity* and (3) *thermal diffusivity*.

HEAT CAPACITY (C) is the amount of thermal energy required to raise or lower a unit mass of the material through one unit of temperature. It is sometimes given as *specific heat* (C) the ratio of the heat capacity of the relevant material compared to that of water.

THERMAL CONDUCTIVITY (λ) is the *rate* at which *heat energy* will transfer through a material relating energy to distance, temperature difference, time and area of cross-section.

This value is constant for a material for a given temperature level, but tends to fall with rise in temperature. The latter effect arises from the interaction of the means of thermal conduction outlined below.

THERMAL DIFFUSIVITY (a) is important when it is necessary to consider the effects of temperature differences set up in a material during transfer of heat. The value relates the conductivity (λ) to the heat capacity C per unit volume.

$$a = \frac{\lambda}{C\rho} \text{ where } \rho \text{ is the density of the material.}$$

A material of high heat capacity will tend to have low diffusivity (i.e. will heat at relatively slow rate) and vice versa. Because any retardation to the overall rate of thermal transfer through a material will increase the local temperature difference when heat is flowing, a high diffusivity is desirable if thermal gradients which could cause overheating or mechanical failure are to be minimised.

Associated with heat change is also a volume change, expansion or contraction related to temperature change by the *coefficient of thermal expansion* which may be measured in linear or cubic units for a solid and in cubic units for a liquid. The coefficient of thermal expansion tends to rise with temperature because a given increase in thermal energy has a proportionally greater effect on the vibration of an atom or molecule as temperature rises and the level of inter-atomic instability is approached. Most materials increase in volume with rising temperatures but there are some exceptions.

Each of these energy relationships depends on the state of the material. Both thermal conductivity and diffusivity cease to have much meaning for the liquid state since in that state the principal mechanism of transfer becomes *convective mixing*. Changes occurring within the solid state also cause associated changes in heat-properties but the latter changes are not usually as great as those associated with change from solid to liquid and vice versa. Change of state is associated with *latent heat of change* which is the absorption or release of thermal energy per unit mass of material as its orbital systems stabilise into their new relationship. Usually extra heat is absorbed if the change is brought about by rising temperature and a corresponding amount released with the falling temperature change, but there are variations from this pattern.

Transfer of thermal energy through a solid material may take place by elastic vibration and by electron transfer. The first will be the major or only mechanism in a solid within which electron mobility is limited and the latter will tend to predominate when common orbital paths are available, notably in metals.

Transfer of heat by elastic vibration is not an efficient mechanism. Atoms and molecules on the surface of a material next to a heat source are caused to vibrate more violently as they absorb thermal energy. These surface atoms and molecules are rebounding elastically to and fro against each other and against adjacent inner atoms and molecules which are not themselves directly exposed to the heat source. The latter are also caused to vibrate more violently by the increasingly frequent random impacts that occur. The inner excited units in turn rebound against more deeply located neighbours and so on. The efficiency of transfer of energy will depend on the relative frequencies and directions of the random impacts that cause spread of the vibrations. These will themselves depend on the configuration of the molecular structure and the types of bonding that prevail. If the surface atoms, which are not so firmly held as inner atoms, vibrate too violently they may be able to break free and the material begins to melt or vaporise. Steep temperature gradients are likely in materials where the vibration mechanism is the predominant means of heat transfer. Such gradients will be associated with stresses generated in the material by the abrupt changes in its dimensions accompanying the steep temperature differences. Materials subjected to this latter condition are said to suffer from *thermal shock* and are liable to mechanical failure from the stresses. If the temperature is varying and giving a fluctuating stress the latter can cause failure by *thermal fatigue*.

Transfer of heat by electron transfer is a comparatively efficient process, but is possible only if there are common valency orbital paths available within the material or if the energy barriers between adjacent orbitals are very low. The greater the number of possible orbital paths the greater the efficiency of transfer. When in a suitable material, heat is absorbed at a surface the valency electrons of the surface atoms are caused to vibrate more violently along with the other atomic particles. Obviously many of these electrons will rebound along any available orbital paths into the interior of the material, carrying the effects of thermal energy with them and transferring it as they interact with other previously unaffected valency electrons and with other ions. Previously unaffected electrons will also be moving to the surface orbitals to refill emptied paths, becoming themselves energised in the process. The nett·effect can be that electron cloud density increases away from the heat source and decreases near to it making the material electropositive near to the heat source, see Figure 4.20, an effect known as the *Thomson* or *Kelvin* effect. Usually the degree of ionisation tends to increase with temperature difference within the material.

The averages of the movements of the electrons in two different materials will differ; therefore, if the materials are in contact with each other, see Figure 4.21 and similarly acted on by a heat source,

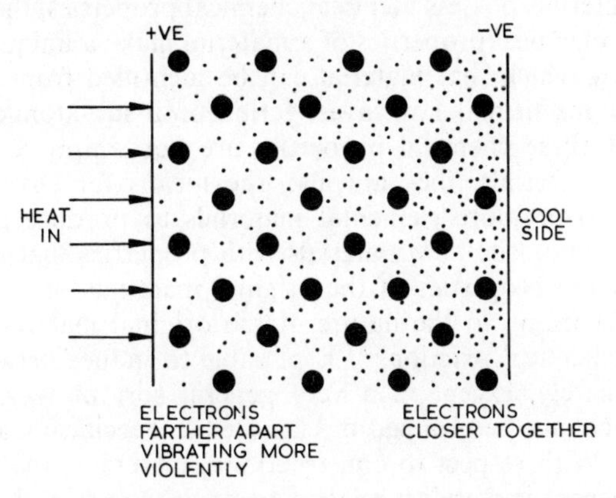

Figure 4.20. Heat transfer by vibration reaction-induced free electron transfer giving rise to a potential difference. Note that the ions (black circles) near heat input also vibrate more violently and transmit vibration onward. Electrons shown as dots

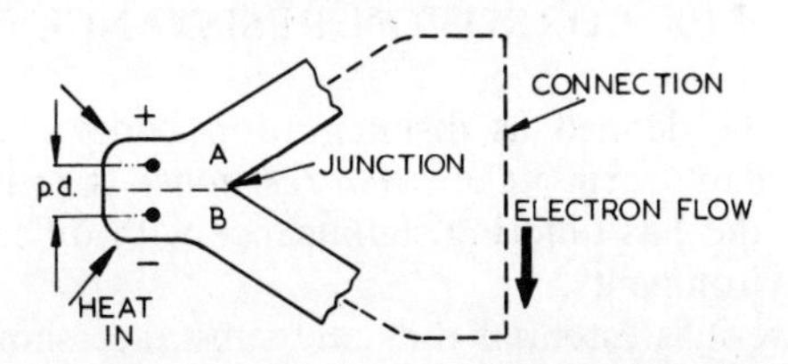

Figure 4.21. A thermocouple formed by heating a suitable junction between different materials A and B. A is assumed electropositive with respect to B

a potential difference will exist between them because they are not ionised to the same level. If an electric circuit is suitably completed a current will flow between the materials (the *Seebeck* effect). With a suitable conjunction of two such materials a *thermocouple* is formed.

Metals tend to be good conductors of heat primarily because there is sharing of valency electrons in their numerous orbital paths and so electron transfer is easy, but elastic vibration transfer also contributes to the total effect.

4.9 CHEMICAL PROPERTIES OF MATERIALS

It is true to say that the chemical properties of a material are only the outward evidences of its inward atomic make-up as revealed by the interaction of its constituent atoms with atoms of other materials. No two materials possess identical chemical properties, therefore the sum of the chemical properties of a material make a unique practical test code by which that material can be identified from all others, without having to make an investigation on a sub-atomic scale.

However these chemical properties are not simply a means for identifying materials, they are also the means for enabling us to combine two or more elemental materials to produce a range of atomically composite new materials with properties that are a compromise of the properties of the original materials or are uniquely different according to the nature of the original materials and the variety of chemical reactions it is possible to induce between them.

Unfortunately, except in a very general sort of way, chemical properties cannot be grouped or classified as specifically as physical properties. With respect to constructional materials, the significant chemical aspect is stability relative to environment and discussion will be confined to that area.

4.10 CORROSION RESISTANCE

Corrosion can be defined as disintegration under chemical attack and can take many forms. *Corrosion resistance* is resistance to this disintegration and has differing significance with differing materials in differing environments.

In every case it is essential that one substance should be within such close proximity to another substance in such a situation that one reacts on the other and absorbs atoms or molecules from it because these atoms or molecules find more stable energy states by making the change. Such detachment can lead to either (1) complete absorption of the attacked substance (2) chemical disintegration of it (3) physical disintegration of it (4) combinations of (2) and (3). Changes of these kinds may occur between two phases present within one material or between two separate materials or between two differing phases present one within one material and one within another material in close proximity to the first material. The worst conditions of disintegration are likely to occur when the reaction products form either a liquid or a loose powder with no coherent strength. Three types of corrosion have to be taken into account:

(a) direct internal reaction
(b) direct external reaction
(c) indirect reaction

4.11 DIRECT INTERNAL CORROSIVE BREAKDOWN

To achieve certain specific properties in a particular material it is sometimes necessary to include within it two or more constituents in unstable relationship with each other. The constituents are brought together knowing that there can be a gradual harmful change in their association but relying on the time delay in the reaction to give the material a sufficiently useful length of service life.

The conditions in which corrosive change of this kind can occur may differ widely with material and environment. A reaction may proceed steadily but at such a slow rate that no one bothers about it, or it may be faster and necessitate replacement after a specified time. Such materials are said to *age* because their properties change with time.

The principal governing factor is usually diffusion within the material of atoms or molecules of one or both of the reacting materials through the barrier of the reaction product, as indicated in Figure 4.22. The thicker the reaction layer becomes the longer the time required for diffusion to take place and the slower the reaction becomes (as shown for a surface film in Figure 4.23a). However a

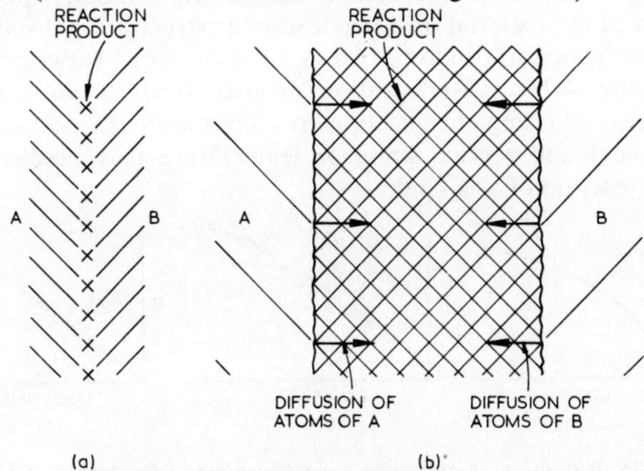

Figure 4.22. Contact reaction between two materials (a) reaction zone (b) rate of reaction and zone of reaction dependent on diffusion rate

reaction need not be total to cause damage to a structure. A small critically distributed reaction product is sometimes all that is needed to spoil desirable properties such as strength, electrical conductivity or thermal conductivity.

Many natural and many synthetic materials are subject to internal chemical changes of this kind even though the effect is usually masked by other changes or is slow. For example, some alloys of metals are used in a *metastable state* in which an unwanted chemical reaction is so greatly retarded by the barrier which the lattice structure presents to diffusion of the reactants that the reaction may take centuries to develop to a critical state.

Although this form of corrosion does not seem currently to be of great importance, its importance is likely to grow as more complex synthetic materials are developed. It may be that it is already a more important factor in some materials than is realised, particularly when such a reaction can be critically accelerated by abnormal changes in use or transient changes in environment. Materials with relatively unordered internal molecular arrangements are likely to be most susceptible to such effects.

4.12 DIRECT EXTERNAL CORROSIVE ATTACK

This form of corrosion is more obvious and more common than direct internal attack since its effects can usually be seen soon after it begins. If a substance known to react with a constituent in a material is present in the environment then reaction will occur. If the reaction is a continuing process, it may seriously modify the whole structure of the material or it may cause the structure to disintegrate locally by penetrating along phases or boundaries between grains (see Figure 4.23). Disintegration exposes fresh surfaces of the material so enabling the reaction to continue further. Changes in environment and service stress or temperature may accelerate or retard a reaction of this kind.

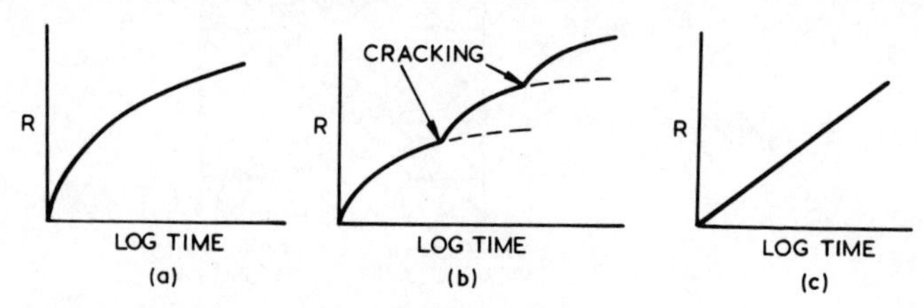

Figure 4.23 Variation of atmospheric corrosion rate with nature of corrosion skin (a) continuous adherent permeable skin (b) fracturing adherent permeable skin (c) non adherent skin

Probably the most obvious reaction of this nature is simple oxidation by the oxygen in the atmosphere with almost every material. Its effect may be no more than to cause the appearance of a patina on a bright surface, or to cause a surface to become dulled, but it can cause gross change in the surface of a material, e.g. by burning. Temperature has a marked effect on oxidation, the reaction usually becoming more rapid with rising temperature until, in suitable materials, ignition or even explosion may occur.

The extent and the progress of oxidation depends on the nature of the oxide film that is formed. If it is adherent it becomes a diffusion barrier tending to slow down the process in the manner described for internal corrosion except that one side of the barrier is open to the atmosphere instead of to another solid. For the latter reason if an oxide film is non adherent or fluid it may peel away leaving a fresh surface exposed to attack. In these circumstances the attack proceeds continuously. If the film is adherent and completely impervious to diffusion oxidation stops until the skin is broken or removed. So called *stainless steel* has this kind of oxide film at normal temperature. Should the film be permeable the oxide attack will slow down with the increasing distance required for diffusion. An oxide film is likely to have a different volume to that of its parent material. Therefore an adherent film will be under increasing biaxial stress as its thickness increases and may eventually crack and permit a fresh acceleration of oxidation as shown in Figure 4.23b. An adherent oxide of greater volume than its parent material will be in compression and much less liable to fracture than one of lesser volume which will be under tension. Thus, an adherent oxide film with a low diffusion rate relative to both its reaction constituents, a slightly larger volume than its parent material and an inherent mechanical strength will soon form into a very effective barrier to continued oxidation and, if mechanically damaged, will soon heal itself.

If corrosion reaction is localised, say around an embedded phase the deterioration can be catastrophically rapid by detachment of phase particles from their matrix; but the final result depends on (1) the rate of the corrosion penetration (2) the size of phase particles and (3) environmental conditions.

Although oxygen attack is an obvious danger and probably the most common form of direct chemical attack there are other attacking agents that can be present in a gaseous atmosphere and there are other situations in which attack can take place. Of the latter the most obvious example is that in which a liquid or other corrosive substance is in contact with a solid surface say in a container. It should be noted that a situation of this kind does not mean, necessarily, that direct reaction is the only possible kind of attack. In fact if the corrosive substance is an electrolyte, such as an aqueous solution in contact with an electrically-conductive solid, the electrolytic form of corrosion, described in Section 4.13 is just as likely to occur.

Concentrated liquid chemicals and liquid metals are likely agencies for direct attack, particularly if the reaction product dissolves away into the liquid keeping fresh reactant solid surfaces continuously exposed. Dissolution or disintegration of a solid exposed to such attack can be very rapid. On the other hand, if the product of reaction is an impervious strong film strongly adherent to the solid, the corrosive attack may speedily slow down to a safe rate or even stop until the film is damaged by some means.

4.13 INDIRECT CORROSIVE ATTACK

In this form of corrosion the reaction is not directly and solely between two adjacent substances but requires some other outside agency. Three types are known.

(1) Electrochemical
(2) Electrolytical
(3) Catalytic
(4) Irradiation assisted

Each type of attack occurs in particular circumstances. The best known and commonest is the form of electrochemical attack that takes place in the moist atmospheres frequently present near a sea coast or in heavily industrialised areas. The other types of attack are not so frequently encountered, but in situations in which they do take place can be just as costly and harmful.

ELECTROCHEMICAL CORROSION takes place as a result of the inherent electrochemical differences existing between differing electrically conductive materials or similary differing phases within materials, causing electric currents to flow in circumstances which permit electrolytical separation or transfer of the atoms or molecules of one of the materials. It should be clearly understood that the latter action must be present before a corrosive attack can take place by this means. The mere completion of an electric circuit and the passage of current does not entail disintegration of a material (see thermocouple Section 4.8). Electrochemical corrosion may also be initiated by a potential difference set up within a single phase material by temperature difference or created by local variations in the environment.

In each case a natural potential difference must exist between two areas of conducting material in close contact with each other and an electrical circuit must be completed through them by way of a suitable *electrolyte*. (An electrolyte is a conductive compound which

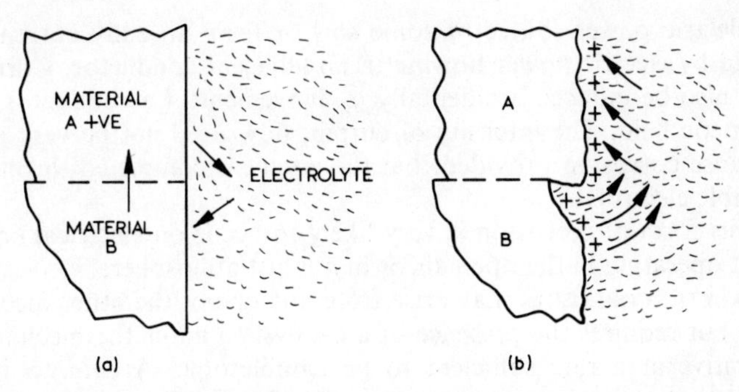

Figure 4.24. Electrochemical corrosion of a metallic material B electronegative with respect to metallic material A. The presence of a reactive electrolyte completes the circuit .(a) Electrons follow path of arrows from B to A internally. Cations released from surface of B travel externally through the electrolyte to A (b) Cations detached from B by reaction with electrolyte erode B and migrate to A

in the liquid state or dissolved in a liquid solution will disintegrate with the passage of an electric current.) A circuit may be completed as shown in Figure 4.24 and the electrolyte as it disintegrates may react with the more electronegative material to detach positively charged ions from it (electrons being drawn directly into the more electropositive material) and transport these ions towards the other material in an effort to balance the electropotential. Progress of the attack may vary with many things. Hydrogen is usually released from the electrolyte and gathers on the surface of the more electropositive material eventually stopping the current (and of course the corrosive attack). Removal of the hydrogen restarts the process. Natural convective circulation in the electrolyte may help remove the hydrogen or the potential differences in the circuit may be enough to keep the current breaking through the barrier.

The form of attack depends on the distribution and nature of the reacting solids. Attack tends to concentrate near to the boundary between the solids therefore the more electronegative solid may have its bonds with the other loosened or it may even become completely detached. Small pieces of the electronegative material may simply be dissolved away. Irregularities on the exposed surfaces or other causes may create differences in concentration of the electrolyte and change the corrosive patterns, in some cases speeding it up locally or, in other cases slowing it down.

ELECTROLYTIC CORROSION takes place in situations in which electrolytic disintegration is caused by *application* of an electric current from a source other than the materials themselves. The mechanism is the same but the electric current is the result of stray current from an electric power source of some sort or from an eddy current induced by electric power flowing in an adjacent conductor. Currents may also be induced accidentally by movement of a conductor in a magnetic field. The intensity of current flow need not be very great to cause corrosion provided that the circuit is completed through a suitable electrolyte.

This form of corrosion is very likely to occur on electrical equipment operating in the open air or in a moist atmosphere.

CATALYTIC CORROSION may arise from any one of the other mechanisms but requires the presence of a catalyst to make the mechanism operative at a rate sufficient to be troublesome. A *catalyst* is an agent that assists a reaction without itself seeming to be affected by the reaction. It is believed that the catalyst acts as an intermediate station, permitting easy exchange movement of diffusing atoms or molecules. There are many possible catalysts and a simple example is the effect of water, or water vapour on a direct reaction such as oxidation.

IRRADIATION-ASSISTED CORROSION can take a number of forms depending on the materials, the environment and type of radiation. In such a case the irradiation of one of the materials leads to electronic or structural changes so that its properties are changed. The change in properties may take the form of deterioration in mechanical strength, leading to disintegration, or it may convert a previously stable material into a reactive form and subject to one of the direct corrosion processes. Some forms of radiation may induce ionisation, change the electropotential balance and generate electrochemical attack.

Infra-red radiation can induce temperature changes that may cause structural change or induce potential differences in irradiated materials. Ultra-violet radiation can change the bonding arrangement within complex molecular materials and cause mechanical disintegration. The hardening and cracking of natural rubber and some synthetic materials in sunlight is an example of this. Other forms of radiation such as γ and X radiations can cause structural changes and induce ionisation. Radiations resulting from thermonuclear reactions can have very marked effects on some materials causing serious internal structural damage and marked structural changes.

BIBLIOGRAPHY

EVERETT, A., *Mitchell's Building Construction—Materials*, Chap. 1, Batsford (1970)

HAYDEN, W., MOFFATT, W. G., and WULFF, J., *Structures and Properties of Materials*, Vol. 3, Wiley

ROSE, R. M., SHEPARD, L. A., and WULFF, J., *Structures and Properties of Materials*, Vol. 4, Wiley

TWEEDDALE, J. G., *Metallurgical Principles for Engineers*, Chaps 11 and 12, Iliffe, London (1962)

5 Ideal and Real Properties

It should now be apparent that there is much that is not known in materials science. There is detailed understanding of the natures of individual atoms but there is less understanding of the behaviour of those same atoms when they are interacting with each other in bulk. In fact some aspects of behaviour seem to change radically in an unpredictable manner from one type of association to another between somewhat similar groups of atoms.

Theoretical predictions about the behaviour of the different types of atoms forecast certain properties in the materials they form but, in practice the properties actually obtained can diverge very widely from the theoretical.

It is only when the nature of these differences is understood that any really scientific approach can be made to the design of solid materials for specific purposes. Knowledge of this subject is still very much in the development stage and the impression of precise knowledge and prediction, conveyed by much of the literature on the subject, is largely an illusion. Where does theoretical prediction diverge from empirical result?

5.1 IDEAL PROPERTIES OF MATERIALS

In considering the structure of atoms and molecules previously it was shown that certain types of bonding are possible between particular types of atoms; but it was pointed out that it is not always possible for the bonding potential to be fully used and the nature of the prevailing bonds may change with time or ambient conditions, perhaps alternating rapidly and unpredictably. It was also indicated that some bonds are strongly directional whereas others are not, some are mechanically strong and some weak, some permit interchange of electrons and some do not. Another factor mentioned was the different sizes of atoms. How are these factors taken into account in determining the possible properties of a material? The properties of any solid material must be a compromise between the following factors

(1) types of bonds that prevail within the material
(2) respective numbers of each type of bond
(3) relative strengths of each type of bond
(4) relative distribution of each type of bond within the mass
(5) relative orientations of bonds within the mass

Knowing the proportions of atoms of differing types present in a material, the characteristics of each type of atom present and the state of the material at the time, it is reasonably practical to predict what the ideal possible combination of properties should be in the circumstances. How realistic can such predictions be?

It is possible by scientific experiment to verify in isolation the accuracy of knowledge of each distinct phenomenon characteristic of specific atoms so the values used to take account of each phenomenon are realistic and accurate. Furthermore, by very careful preparation of material it is experimentally possible to produce, in very small quantities, some materials which do give very nearly the ideal values predicted by theory. Therefore it can be accepted that the development of the ideal properties in a material is potentially possible. Also it is now known why ideal properties are very rarely obtained from a commercially produced material.

5.2 DEFECTIVE NATURE OF REAL MATERIALS

Whenever large numbers of atoms are brought together to form a solid material there are divergences from uniformly orderly behaviour; there are defects in the structural arrangement. Even the most orderly materials, the crystalline materials, do not have anything like complete uniformity in their macrostructure. However these obvious divergences are not the only ones that can appear and all materials contain many smaller-scale defects within their microstructure. For example, all the atoms or molecules within a material structure are not equally able to complete the optimum number of bonds possible. Therefore there will always be irregularities in the arrangements of groups of molecules and in the structures of individual crystals. In fact it may be said with truth that materials science is based on the understanding of defects in materials and that materials technology is the use and control of these defects.

The differences between ideal and real properties are not all undesirable since in some cases it is this difference that makes it economically practical to produce materials in a usable form. A notable example of this is the ease with which many metallic materials may be manipulated into useful shapes.

5.3 MICROSTRUCTURAL DEFECTS

Microstructural defects are associated with the atomic and/or molecular configuration within the structure of a material and may be defined as irregularities or abnormal variations in the pattern of the bonding and of the distribution of the atoms or molecules making up a microstructure. Accumulations of such defects are always present in the vicinity of macrostructural defects and in some cases themselves can total up to a macrostructural defect. The relative importance of each type of basic defect varies relative to the nature of the structural features within a material. The degree of order within a structure may determine what is a noteworthy defect; a disordered amorphous structure is full of irregularities in bonding arrangements and distribution of atoms, implying that the material is mainly composed of defects and that no one type is of particular importance. On the other hand, if a structure is very orderly quite small differences within it can be very important. It is usual to group microstructural defects into three classes

(1) point defects
(2) line defects
(3) surface defects

POINT DEFECTS are defects in which there is an anomaly in the arrangement of atoms within, or about, a basic molecule or within a cell of a crystal. The defect may take one of three forms, namely a *vacancy*, a *substitution* or an *interstitial addition*.

A *vacancy* exists when a single atom or molecule is absent from the position it should occupy and there is a hole in the structure.

A *substitutional defect* occurs when an atom or molecule of one kind occupies a position normally occupied by one of another kind.

An *interstitial defect* is present when an extra atom either of the same kind as those already present, or of a completely different kind, is present in a position not normally occupied in that structure.

Such defects as these cause a local structural distortion, the amount depending on the circumstances but the effect is unlikely to be very noticeable. As the numbers of such defects increase, however, they may have very marked effects on properties.

LINE DEFECTS form a group of defects of particular importance to crystalline materials such as metals. This kind of defect is formed when a row of atoms or molecules is missing or is misorientated in relation to the rest of the structure. The first situation is not so common as the second.

If a row of atoms is missing the surrounding structure is likely to distort a little but the only other effect is to provide a tunnel through which diffusion could take place. It is not likely that many rows will be missing near to each other, because a structure can usually adjust itself to absorb them by self-diffusion. However, many rows may be missing on or near severely distorted boundaries when interaction with adjacent phases prevents adjustment to absorb them.

Misorientation of a line of atoms or molecules in the form of a *dislocation* is very common in crystals, particularly in metallic materials, and can have important effects on properties, notably the mechanical properties. A dislocation is a continuous line marking the boundary, in an otherwise ordered structure, between an area on one side of the line which is linked in an orderly array but is one space out of register with the similar orderly array of linking in the complementary area on the other side of the line (Figure 5.1). Along

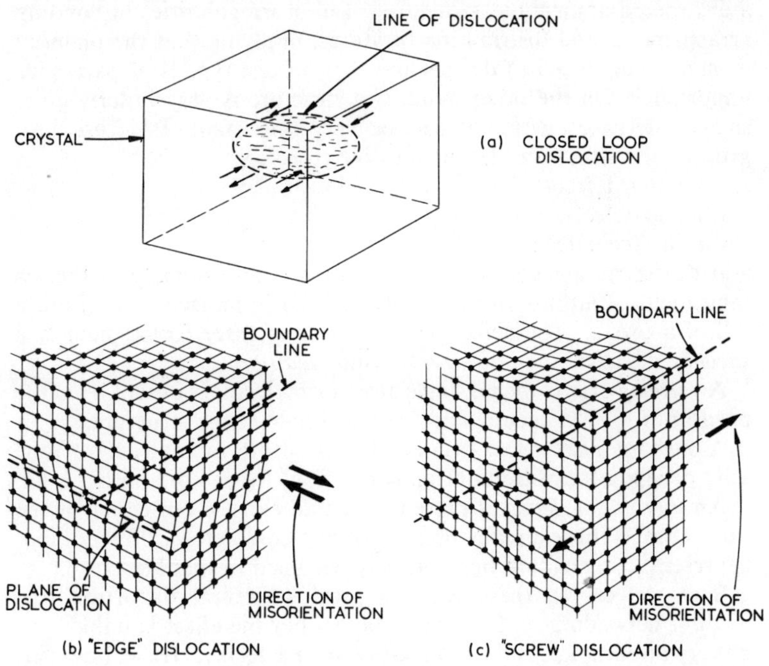

Figure 5.1. A dislocation such as shown in (*a*) *is made up of components of* (*b*) *and* (*c*)

the line there is a succession of atoms or molecules which on one side are linked into the main pattern of the system but on the other side are suspended between two alternative links, one belonging to the main system and the other belonging to the out-of-register system.

It should be noted that because dislocations are very difficult to represent accurately in diagrams, the convention of representing them only as in simple cubic systems is normally used, although this gives a very restricted picture of the conditions.

A dislocation is made up of two types of mislinking—an *edge* type and a *screw* type see Figure 5.1—which by combining in differing numbers and sequences and opposite handing, can adapt the dislocation to balance whatever influence is causing the disturbance. Note that the displacement caused by a screw lies along the line so it moves in a direction normal to the direction of displacement. The cause of a dislocation is a shearing strain in the lattice resulting from the presence of other defects such as dislocations orientated in other directions, impurities, grain boundaries or from applied stress. Compared with other defects a dislocation is mobile within its parent lattice system and is particularly sensitive to shearing stress applied within its plane(s) of displacement. Dislocations are

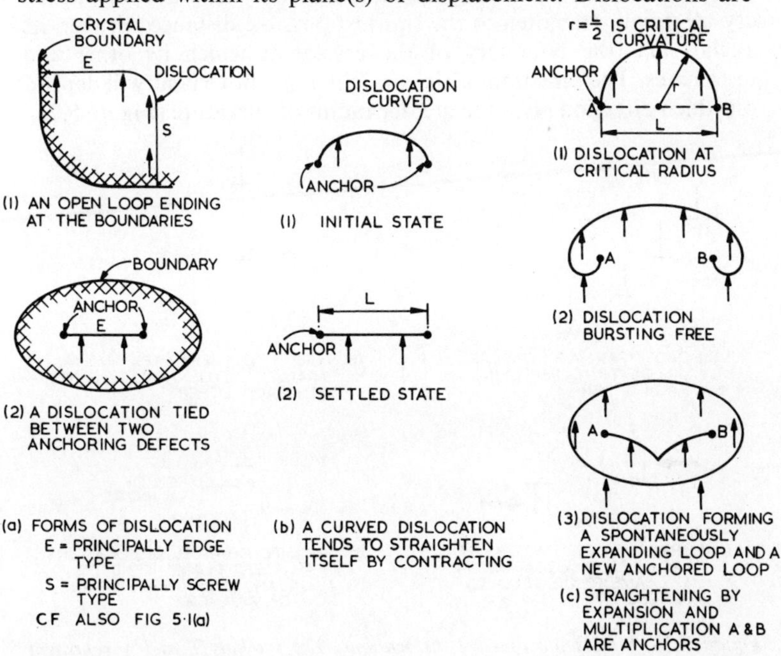

Figure 5.2. Dislocation behaviour. Small arrows show direction of misorientation

present in all normal crystalline materials and more dislocations can usually be generated by suitably applied stress. All dislocations must conform to certain simple patterns of behaviour if they are to fit into the energy balance of a crystal system. Departure from these patterns requires such high energy changes in the form of strain in the lattice and/or the creation of other defects that it does not occur except in unusual circumstances and, when it does occur, is likely to lead to mechanical failure of the material. Some of the patterns of behaviour are outlined below.

A dislocation must be continuous within a crystal, that is it must form a closed loop or end at a crystal boundary or at another intersecting dislocation of different orientation of displacement (Figure 5.2a). The plane of action of a dislocation within a crystal may change direction provided that the amount and direction of misorientation of the lattice is unchanged.

A dislocation must always try to straighten itself because the more sharply it is curved the greater is the elastic strain in the system. If the dislocation is a closed loop its line will expand to the grain boundaries, or if it is an open loop anchored at both its ends, it will settle into the best straight line between its anchor points (Figure 5.2b). If it is an open loop anchored at only one end inside the crystal it will straighten to the shortest possible distance between its anchor and the boundary of the crystal at which its other end terminates. The direction of the straightening movement will depend on which direction gives the greater radius of curvature (Figure 5.2c).

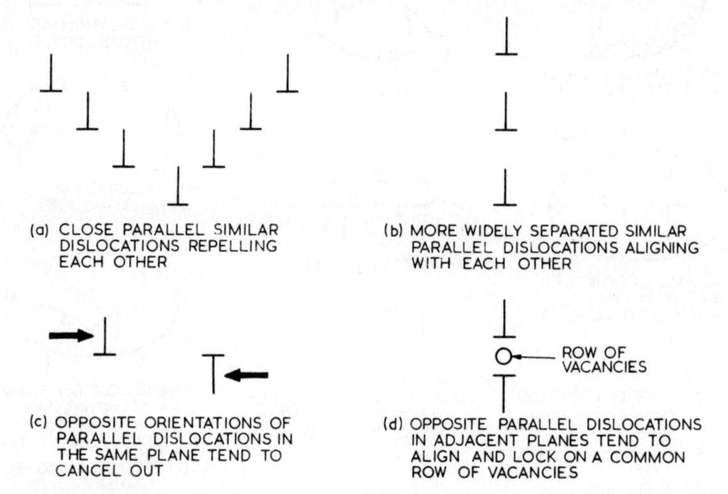

Figure 5.3. Interaction of edge dislocations. The symbols T *and* ⊥ *represent the orientation of the extra part-plane of atoms*

If more than one dislocation is present in the same or closely adjacent planes they will tend to keep away from each other if their associated distortion patterns are of similar orientation (Figure 5.3a).

If the distortion patterns are of opposite orientation and in the same plane they will tend to come together and cancel out. If they are one plane apart, they will tend to come together and leave a line of vacancies (Figure 5.3d). In the latter case they lock together (forming a *sessile* dislocation) and become very difficult to separate.

If dislocations of similar orientation of distortion pattern are separated by a sufficient number of planes of the lattice, they tend to settle in alignment above each other so that their total distortion strain is kept to its minimum (Figure 5.3b).

If a dislocation moves right through a crystal it leaves ledges (Figure 5.4a) of opposite orientation one at each end of the crystal

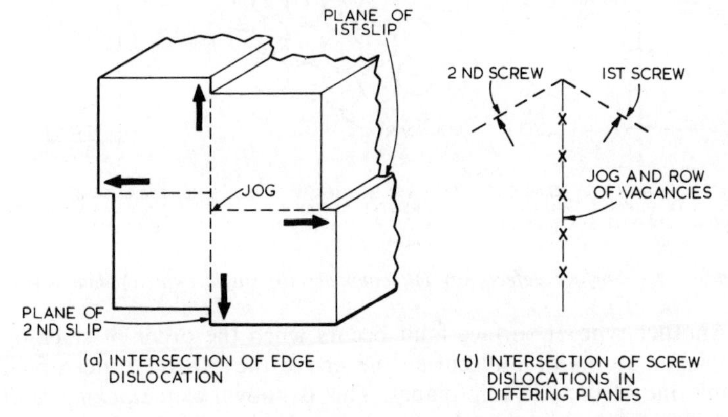

Figure 5.4. Intersection of dislocations travelling in different intersecting planes

on the plane of, and in the direction of misorientation.

Dislocations moving in differing directions in intersecting planes cause a *jog* or step in the continuity of the lattice bonding if they intersect (Figure 5.4a). Screws as they intersect, leave a jog and a regular series of vacancies along the line of their intersection (Figure 5.4b). Jogs may make further dislocation movements in the same planes more difficult and vacancies certainly interfere with movement.

It is because dislocations can influence many properties to such an extent that they are considered here in more detail than are other defects.

SURFACE DEFECTS occur when a block of planes making up part of a crystal or other similar orderly structure is terminated or is out of alignment with adjacent blocks along an internal surface plane.

An outer surface is an obvious example of this kind of defect but a low-angle sub-grain boundary misalignment can also come within this group as, for example when a series of edge dislocations are stacked uniformly in alignment one above the other within a crystal as shown in Figure 5.5a to give a *tilt boundary*.

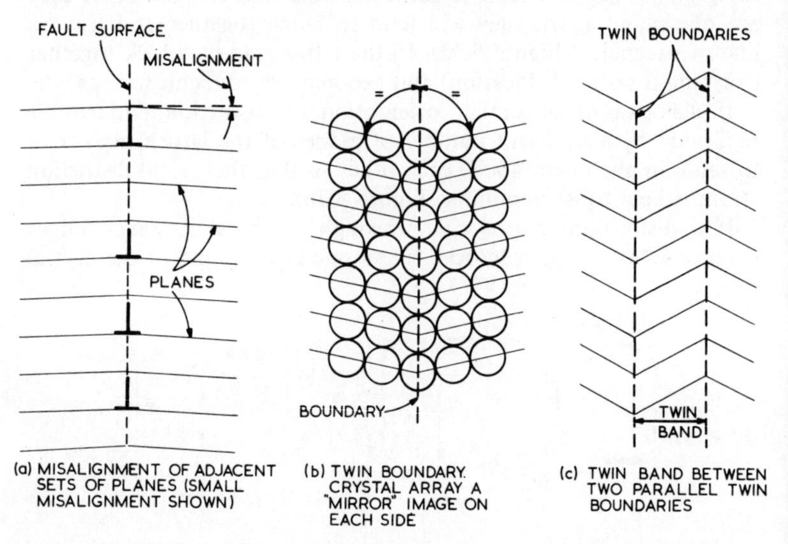

Figure 5.5. Surface defects (a) Tilt boundary (b) single twin (c) double twin

Another type of surface fault occurs when the order of stacking of planes atoms or molecules, one above the other is interrupted along one of the stacking planes. This is known as a *stacking fault* and may arise in any stacking system in which planes of atoms or molecules lying in geometric arrangement relative to each other have alternative relatively stable alignment positions for stacking against each other. The best known example can occur with close-packed planes. Normally, a particular type of material whose planes pack in this way will stack with either every alternate layer aligning (hexagonal close packed system) or with every third layer aligning (face centred cubic system); but it can happen say by partial slipping under particular conditions, for the stacking to be interrupted by the alternative stacking location being taken up. In this case instead of alignment being in the order of ABABAB it may become locally ABCABA or if the packing is ABCABCABC it may become locally ABCABABC.

A particular series of stacking faults gives the surface defect known as a *twin boundary*. In this case one part of an orderly stack is restacked, usually by local slipping so that although the same stacking system prevails the stacking orientation is changed across the fault (Figure 5.5b) one part being a mirror image of the other. Very often two such faults of opposite hand occur fairly close together keeping the outer parts stacked in the same alignment but separating them by a differently orientated *twin band.*

5.4 MACROSTRUCTURAL DEFECTS

There are four types of macrostructural defects

(1) misorientation of adjacent molecular groups or crystals
(2) boundary defects between adjacent groups
(3) gross impurities
(4) cavities

MISORIENTATION OF A STRUCTURAL GROUP is a defect almost certain to be present in a greater or lesser degree with every material. Even crystalline structure in the form of one large single crystal may contain locally misorientated groups within itself although the degree of misorientation is likely to be small, perhaps less than 10° and attributable to the presence of large numbers of microstructural defects grouped in particular ways. Such a crystal is said to contain *sub-grains* (Figure 5.6). Also, a seemingly perfectly arrayed large single crystal must contain a random arrangement of magnetic

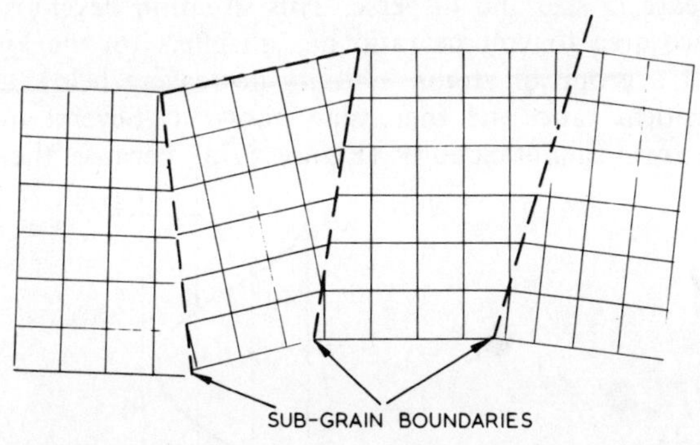

Figure 5.6. Small angle misorientations within a main grain giving a subgrain structure

domains (Section 4.7) if its overall magnetic field is to balance within itself and these domains may be associated with dimensional differences within the crystal lattice. Misorientation can arise from a number of causes. The most common cause is the mode of solidification.

The atoms or molecules in a solidifying liquid, usually under the energy drive set up by falling temperature, have to stabilise their positions relative to each other before they can settle into the solid state. If the bonding characteristics of the constituent atoms or molecules are not suited to forming a regular crystalline type of array the atoms or molecules will move less and less randomly against each other and will gradually settle into the best relatively-fixed association they can achieve in the circumstances. This kind of association may involve short-range ordering in local groups or grains but the average arrangement will appear amorphous. Sizes and shapes of groups will vary and some local groups will be more stable than others but a state of complete stability is unlikely.

When atoms or molecules are able to unite in an orderly array, they are often readily solidified into a mass from a thermally-induced molten state of random movement, by cooling down through the solidification temperature. Ordering is likely to take place in a number of zones simultaneously, each centred around a stabilised nucleus which is likely to form the core of a distinguishable grain in the solidified material. A *solidification nucleus* is an ordered group of atoms or molecules that forms spontaneously and becomes progressively more stable as units are added to it in a systematic fashion; that is, under solidification conditions a nucleus tends to grow. A nucleus must be of a certain minimum ordered size before it will grow. If it is less than the minimum critical size the chances are it will decrease in size and disperse. This situation develops because the surface area to volume ratio has an effect on the kinetics of growth of a group of atoms, stability decreasing below a characteristic critical ratio and increasing above it. Several nuclei are likely to form simultaneously (Figure 5.7a) because the random

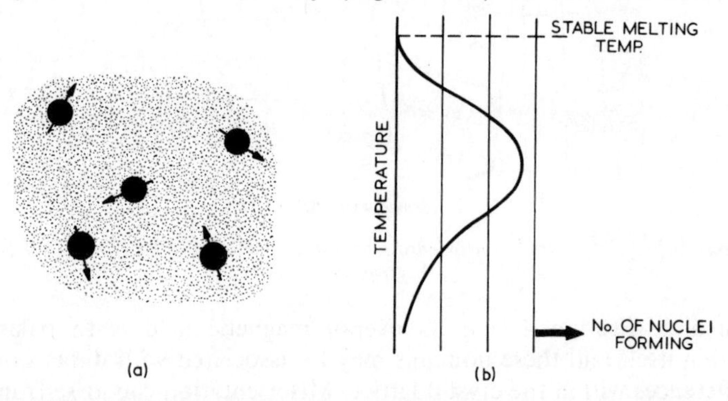

Figure 5.7. (a) Nuclei forming. Arrows denote orientation not direction of growth (b) chances of nucleation from liquid at temperatures below the stable melting and freezing temperature

movements of the units in a liquid phase are likely to be equally random in a comparatively large volume of liquid. Therefore the chances of a minimum nucleus group occurring spontaneously are equally good in every similar part of the volume. The nuclei that do form will tend to grow more or less at the same rate as each other if conditions are uniform, but they will be differently orientated. Hence when their growth faces meet and try to bond the match may not be exact and the nuclei will be held in misalignment. The chances of nuclei forming in a cooling liquid tend to become greater with decreasing temperature as movement slows and the critical size of stable nuclei decreases (Figure 5.7b); but beyond a certain limit, reduced mobility (slower diffusion) begins to hinder rather than to help the chances of ordering. In general, the nucleation of perfectly pure materials is slow to develop (although reaction with container walls may speed the process) and grain size is large. Impurities in a material can help to initiate nuclei and so reduce grain size by creating more numerous solidification centres.

Similar principles of nucleation operate when solidification is occurring by chemical reaction between mixed but unbonded constituents. The driving energy in that case will come from the reaction and not from thermal differences between the mixture and its environment.

BOUNDARY DEFECTS always exist where there is a grain boundary either because there is misorientation between adjacent ordered volumes of the same phase, or because there is a change from one phase to another within the material, or because there is a *free* surface (a surface exposed to the atmosphere or completely unbonded to its next neighbour). Boundary states may vary widely between the following possible conditions

(1) minor misorientation of bonding associated with slightly increased local instability of the intermediate atoms or molecules between similar adjacent volumes within one phase of a material (cf. Figure 3.18a)
(2) marked disorientation between similar adjacent volumes of one phase of a material outlined by a narrow disordered, unstable zone of parent atoms or molecules (cf. Figure 3.18b)
(3) a transition from one phase to another through a narrow zone of gradually changing composition and geometric orientation to blend the two parent phases
(4) a narrow band between two different phases made up by a third material a compound of the first two
(5) complete incompatability between two adjacent phases with little or no bonding between them, the boundary being an almost continuous fissure

(6) combinations of two or more of the other five.

In every case a boundary is associated with less ordered bonding and less systematic ordering than in the main volumes of the material and concentration of any insoluble impurities or surplus constituents that may be present. It is these factors that may tend to make corrosive attack, when it occurs, more severe in grain boundary zones.

GROSS IMPURITIES are groups of insoluble impurities, usually large enough to be distinguishable as phases differing in nature from the phases of the parent material. This is not to imply that there is no other form of impurity present, but only that a particular impurity is so insoluble in the matrix and is present in such quantity that it has formed into a separate entity. Commonly such impurity phases appear at grain boundaries and particularly at meeting points of grain boundaries, as shown in Figure 5.8. The nature, shape and

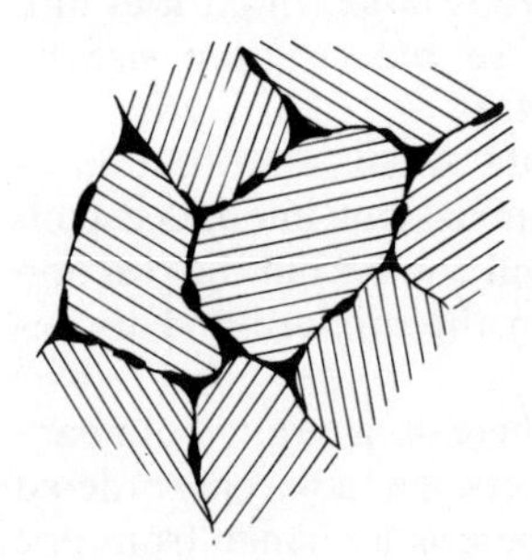

Figure 5.8. Insoluble impurities precipitating at grain boundaries

distribution of the impurity can each greatly influence its effect on properties, the worst situation being likely when it forms a continuous or nearly continuous film along all the grain boundaries. In this situation a small amount of impurity, perhaps difficult to detect, can have disastrous results by preventing bonding across the interfaces or by forming relatively weak bonds with the parent phases.

CAVITY DEFECTS surrounded by exposed surfaces, are often present and can vary in size, distribution and shape with the cause of their presence. They may be left by shrinkage of liquid as solidification takes place; usually rather angular in shape with rough surfaces. Alternatively, gas may be entrapped during solidification leaving either a honeycomb of fairly regular spherical-shaped small pores or larger, perhaps more angular, irregularly spaced smooth walled cavities. Another possibility is that parts of badly bonded boundaries may open up during solidification shrinkage, perhaps assisted by accumulation of gas, to form small lamellar cavities called *microfissures*. Plastic deformation after solidification may change the shape and distribution of cavities that are already present. In some cases cavities may close up and their walls become bonded; in other cases the cavity may extend by flattening and/or cracking.

5.5 THE INFLUENCE OF DEFECTS

If defects are responsible for the differences between the ideal and real properties of metals, it is desirable to know what kind of effects they can have and how these effects can be controlled either to limit or to make use of them. The first aspect is considered in the remainder of this section and the second is included in Chapter 6.

For the present purposes the main features of the principal classes of solid structures are reviewed and then properties are outlined, first in relation to the ideal and then in relation to the influences of defects. Mechanical properties, because of their importance are considered in rather more detail than other properties.

5.6 TYPES OF SOLID STRUCTURES

An outline is given in Chapter 3 of the possibilities that exist for the combination of atoms and molecules within solid materials. The variety of possibilities is almost infinite but the macrostructures of solid materials are usually built up from certain typical basic group associations of the particles (grains) of the phase or phases in the material.

There are five typical groupings that have characteristics sufficiently different to be recognisable and to have considerable influence on the properties of a material.

(a) randomly located, randomly orientated combinations of randomly sized and randomly shaped groups of atoms (grains) held together by a combination of intergranular bonding and geometric interlocking
(b) systematically located, randomly orientated combinations of randomly sized and shaped grains held together by intergranular bonding and geometrical interlocking
(c) randomly located, randomly orientated combinations of relatively uniformly sized and shaped grains (equiaxial grains) held together either by intergranular bonding or by a combination of intergranular bonding and geometric interlocking if the grain shape permits it
(d) systematically located, randomly orientated, controlled shapes and sizes of grains held together by intergranular bonding or by intergranular bonding and geometric interlocking, if the grain shape permits it

(e) systematically located and orientated controlled shapes and sizes of grains held together by intergranular bonding or by intergranular bonding and geometric interlocking, if the grain shape permits it

Greater complexity is likely in a multiphase material because the two or more phases that exist side by side, in proportion to their respective volumes, are unlikely each to conform to the same type of grouping; therefore mixed types of grouping are likely. Most phases within materials tend to conform to (d) although some conform to type (c) and some to type (e).

Important characteristics of each type not already covered are outlined below.

TYPE (a) STRUCTURAL GROUPING is unlikely to give maximum density in a structure because there will be many gaps between randomly orientated grains. For the same reason nothing like the maximum number of intergranular bonds possible will be completed and this will have a proportionate effect on all properties. This kind of grouping is unlikely to give either uniform properties or directional properties.

TYPE (b) STRUCTURAL GROUPING will give greater relative density than group (a) because systematisation of the grain distribution will fill many gaps that might otherwise exist between grains, compacting the structure and making possible both the formation of more intergranular bonds and some systemisation of geometrical interlocking. The increased numbers of bonds will influence the properties of the mass which will become more uniform and may become slightly directional depending on the relative distribution of the grains.

TYPE (c) STRUCTURAL GROUPING will tend to give properties intermediate between those of types (a) and (b). The more uniform shapes and sizes of the grains will permit a more orderly packing of the grains against each other, therefore the intergranular bonding distribution will be more uniform but, depending on the particular size and/or shape of the grains, the bonds may not be as frequent as they might otherwise be and density may be low. Some properties, particularly the mechanical properties, may be directional if the grain shape so determines and with certain configurations of multiphase structures, geometric interlocking can be an important bonding mechanism. Multiphase materials will not give uniform properties if phase distribution is significantly irregular.

TYPE (d) STRUCTURAL GROUPING, since it is controlled both in size and distribution of grains, can give maximum relative density, maximum utilisation of intergranular bonding and optimum geometric interlocking of grains. Properties are more controllable, can be made either uniform or more specifically directional and are potentially nearer to the optimum. In certain types of structure, particularly when intergranular bond strength is low, geometric interlocking can become a major factor in developing mechanical strength.

TYPE (e) STRUCTURAL GROUPING is in many ways similar to type (d) but is more difficult to achieve in most situations. Depending on the type of grain and the intergranular bonding this type of arrangement can be strongly directional in many or all of its susceptible properties. This may be a desirable feature in some situations but more often is not so.

It is desirable to discuss the possibilities of type (d) groupings because of their importance. Five kinds of subgroup are distinguishable (1) basic crystalline (2) molecular crystalline (3) chain molecular (4) rigid network and (5) rigid gel. Each of these is outlined below.

(1) BASIC CRYSTALLINE STRUCTURES are those in which the lattice positions of the crystal structure are occupied by individual atoms. Hence, within a grain, every atom has exactly the same pattern of surrounding neighbours. This does not imply that the atoms can only be of one type, but if more than one type is present, they must be distributed in strict geometric arrangement relative to each other (Figure 3.8). Alternatively, molecules of a suitable shape and combination of atoms could pack together in such a manner that the requirement for a basic crystal is satisfied. This is the case in Figure 3.8 when the basic unit is a molecule of one atom of Sodium with one of Chlorine and that molecules of this kind are suitably stacked in an orderly manner under the influence of coulombic attraction. The main feature of this general type of structure is the orderliness of the placing of the atoms.

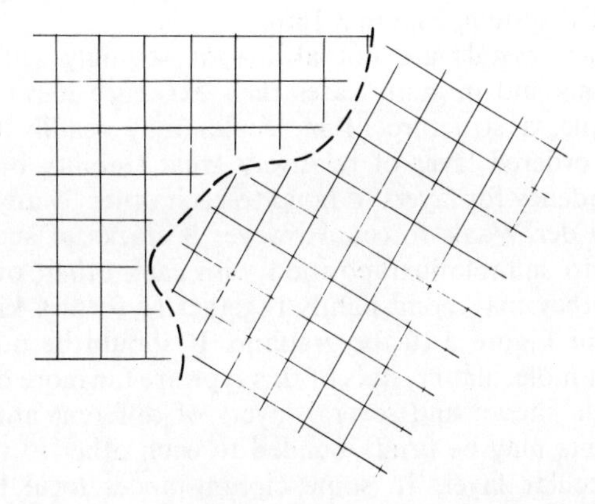

Figure 5.9. Misorientation at crystal boundaries

Direct bonding between individual crystals (grains) of this kind can be relatively strong with nearly all of the potential linkages taken up (notably when the angle of misorientation between grains is low). This bonding is helped by some degree of geometric interlocking (boundaries are never perfectly flat where multitudes of complex crystallographic steps on either side of the boundaries mesh with each other) as shown in Figure 5.9.

When the constituent atoms of a material are all roughly similar in size and the predominant bonding between them is suitable, a systematic crystalline arrangement gives the highest density of packing of the atoms possible in the circumstances. On the other hand, if two kinds of atoms of widely differing size but potentially compatible bonding can intermesh so that the large atoms form a closely-packed crystal structure, then it may be possible for the much smaller atoms to take up certain systematic positions in the gaps in the parent lattice, without causing great disturbance to make up their own super-lattice. Such an arrangement can give more dense packing of atoms than may be possible with the larger atoms alone.

(2) MOLECULAR CRYSTALLINE STRUCTURES differ from the basic crystalline structure in that their lattice sites are no longer all occupied by individual atoms (although some may be) but to maintain the characteristics of a crystal lattice some regularly positioned sites are each occupied by a complete molecule centred on the site, each such molecule being regarded as a unit constituent of the lattice. This is likely to mean that if more than one kind of atom goes into the makeup of each molecule and these atoms are not held in suitably symmetrical positions relative to the molecule centre, then *individual* atoms of one type within the whole structure will no longer each have an identical pattern of surrounding neighbours (e.g. an HCP system, Figure 3.16b).

Molecular crystals are not always as strongly self-coherent as basic crystals and in many cases their existence is rather artificial. For example, a structure of molecules may readily form a geometrically ordered layer of relatively great stability but there may be little tendency for layers to bond to each other by anything except simple van der Waals forces. However if stacks of such layers are brought into suitable juxtaposition with each other, one on top of the other, they may bond lightly together to form a kind of lattice as shown in Figure 5.10 for graphite. It should be noted that the majority of molecular crystals of this type are far more complex than the example shown and several layers of different atoms or atom arrangements may be firmly bonded to each other to make up one basic molecular layer. In some circumstances local bridge-bonds may be formed between layers by using an interleaved layer of suit-

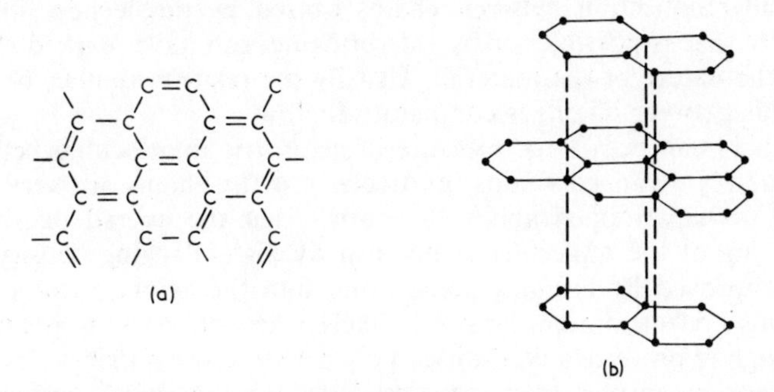

Figure 5.10. (a) Carbon atoms linking to form a stable plate of atoms. The double bonds constantly change position. (b) Graphite formed by alternate alignment of carbon plates under van der Waals attraction between plates

able bridge-forming atoms. Another type of artificial molecular crystal may be made up from long chain molecules when such chains are systematically stacked alongside each other, as indicated in Figure 5.11, and held together either by their van der Waals

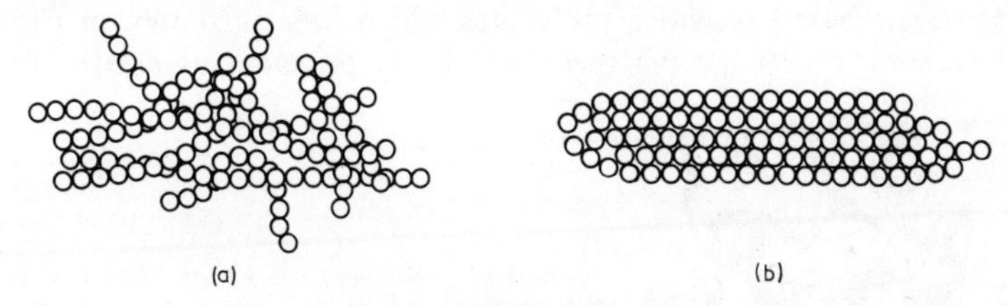

Figure 5.11. (a) Normal randomly tangled chain molecules (b) chain molecules 'crystallised' by alignment

forces or by systematically spaced bridge-bonds formed between them by introducing a dispersion of suitable bridge-forming atoms or small molecules. Such crystallising mechanisms as these are the ones prevailing in *crystallised polymers*, see below. Straighter and more uniform chains crystallise more readily than kinked or irregular chains.

(3) CHAIN MOLECULAR STRUCTURES, as the name implies, are structures formed by the building up of chains of molecules. If a *monomer* is of a type that tends to bond approximately in line with not more than two of its own kind then such monomers may be linked up to form chain polymers. Such chains can be very long and, as they are usually formed in a random manner from a suitable accumulation of monomers, are normally closely and randomly intertwined with

each other rather like the fibres in surgical cotton wool. The degree of interconnection between chains caused by direct side linking, where that is possible, or by side bridging, can have marked effects on the nature of the material. Usually the relative number of such bonds between chains is comparatively low.

There can be a large measure of geometric interlocking between randomly arranged chains, particularly if the chains are very long and perhaps looped about each other, but the overall density of packing of the molecules is not usually high. Packing density can be improved by bringing some order into the arrangement of the chains and maximum density is likely when the packing is orderly enough to produce a crystallised polymer structure if that is possible. A wide range of synthetic materials (notably plastics) conform to one or other of the chain-molecular systems.

(4) RIGID NETWORK STRUCTURES form whenever individual molecules can each join up with three more of their own kind to form a continuous but relatively random stable network or when chain polymers are cross linked sufficiently frequently to produce the same result. Certain configurations of molecules which each form two strongly directional bonds may also form networks.

The most common types of network arrays are found in the glass materials based on silica molecules which link as shown in Figure 5.12, and in *ceramic* materials which are prepared as plastic clays,

Figure 5.12. Random linking of silica molecules (Figure 3.1) to form a rigid network glass

then fired (heated at high temperature) to create a rigid network structure.

A rigid network does not usually have its molecules as densely packed together as they might be if the mode of bonding was different. Limited short-range ordering of the molecules relative to each other may be seen in a rigid network, as determined by the mode of bonding, but the overall effect in the structure is one of random arrangement. Foreign atoms or molecules may often be able to fit into suitable interstices in a rigid network, raising the density and perhaps radically changing the properties of the parent material. Such materials tend to be brittle, because fracture can start easily and spread rapidly.

Some materials form their interlocking bonds only when their temperature falls below a level called the *glass transition temperature*, and reverse the process when temperature rises above that level. Thus natural rubber is softly elastic at n.t.p. but brittle hard at temperatures below 213K.

(5) RIGID GEL STRUCTURES are a special case of network structure in which an open, perhaps not very stiff, network has its interstices filled with another, perhaps more rigid, material or a liquid so that the materials do not tend to separate out and the total structure is more rigid than either material alone. The best known example is cement in which the separate individually strong sand particles are firmly enclosed in a matrix of a complex rigid hydrated calcium silicate (Portland cement) gel network formed by water molecules adsorbing on to and between outer surfaces of the silicate particle layers. (*Adsorption* takes place when atoms or molecules of one material attach themselves firmly to the surface of another material without diffusing into it.)

5.7 CORRELATION BETWEEN TYPES OF STRUCTURE AND PROPERTIES

The marked influence of structure on the properties of a material is an outcome of five features characteristic of each grouping.

(1) types of bonds operating within and between grains
(2) relative densities of bonds of different kinds (the density of bonding is the average number of a particular type of bond operating within a representative unit volume of the material)
(3) relative distribution of each type of bond within a truly representative unit volume of the material
(4) proportionate relative orientation of each type of bond within the representative unit volume of the material
(5) nature of exposed surfaces of the material

Each of these features has its own influence on each of the properties possessed by the material. Some of those influences are clear but others are more obscure so a brief outline is given below.

(1) THE TYPES OF BONDS present in a material are the primary factors controlling the kinds of properties developed in that material (see Chapter 3). One particular type of bond may have a marked effect on a property and conversely another type of bond may have little or no effect on the same property. Each potential property of a material has to be considered first of all in the light of the particular types of bonds that are present.

(2) RELATIVE DENSITIES OF BONDS must be taken into account in

conjunction with the types of bonds because it is the combination of these factors that builds up a property. Thus, a particular kind of bond may offer a weak individual contribution to a property, but if the bond is present in a high density it may become the major influence contributing to that property in a particular material. Conversely, a strongly influential but very low density bond may contribute almost nothing to that property in the same material. Density of bonding is closely related to the relative sizes of the atoms making up a structure.

(3) RELATIVE DISTRIBUTIONS OF BONDS within a material may greatly influence the degree of uniformity and/or directionality of appropriate properties within the material. The density of bonding is an average value over a representative volume of a material, but the averaged bonds may in fact be highly concentrated in one zone and non existent in another. It is obvious both that the types and densities of bonds between grains or polymers will be different from the types and densities within the grains or polymers, and that the shapes and relative volumes of grains or polymers can radically influence the distributions of bonds within a material. In addition to this, in many materials the types of bonds may have quite steady average density values but the bond types may be varying from instant to instant within particular microstructural configurations (e.g. in the graphite structure shown in Figure 5.10, the double bonds shown between some of the carbon atoms although constant in number are rapidly switching between potential pairs within the pattern). Each factor can have its own effect in particular materials in relation to particular properties.

(4) PROPORTIONATE RELATIVE ORIENTATION of bonding is logically the feature next in importance to simple relative distributions of bonds. Every bond in a solid material, even when the bond is not inherently a directional one, must have a directional characteristic because each bond is operating between approximately fixed centres of action dictated by the respective centres of attraction and repulsion of the atoms in each group. If there is overall ordering of the directions of a type of bond within a material, then a property influenced by that bonding will also show directionality. A number of clear examples of this can be found in the magnetic properties of a material but similarly directional effects may operate in many materials in relation to other properties particularly with respect to strength.

(5) THE NATURE OF EXPOSED SURFACES of a material can have a marked effect on many properties. An exposed surface is a region of more marked disturbances to the continuity of the structure than any other kind of surface. For example, intergranular or inter-

molecular surfaces usually show a considerable degree of reaction between themselves, which helps to take up some of the bonding energy available on each surface but an open surface, or a completely unbonded interface, is not stabilised at all unless it reacts with whatever environment is available to form a protective layer. Chemical properties are the ones most obviously affected by the state of an exposed surface, but almost every other property may also be affected to a certain degree.

These different influences can be taken into account by making appropriate statistical analyses of the distribution of bonds in relation to the orientation of a representative unit cube of the material and correlating these with the various properties of the material: the relative orientation of the structure; the overall shape and the finish of the surface as used in a particular situation.

A unit cube is used because it gives a uniform equal cross-sectional shape over a known length on each of its three principal axes and its axes are readily correlated with identifiable structural features; the unit cube should be orientated in a known way in relation to a suitable characteristic feature of the microstructure or macrostructure. Selection of orientation direction is not important if there is no significant degree of directionality present in the material. The unit cube can be considered in relation to each relevant property in turn (see Figures 4.2 and 4.3).

A cube such as that shown in Figure 5.13 has three principal axes.

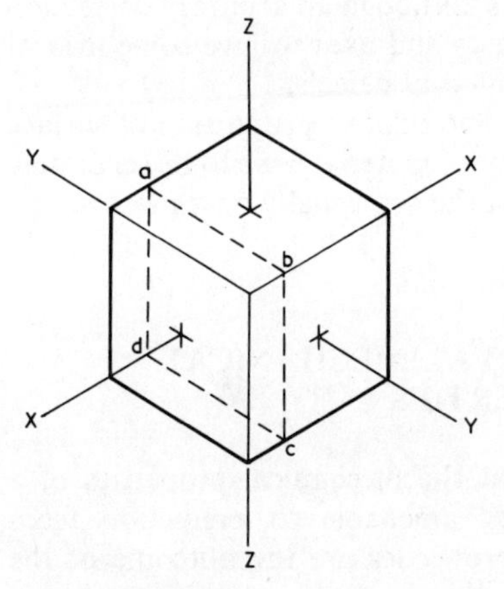

Figure 5.13. Unit cube to identify and correlate anisotropic properties of a material

The properties as measured on any one axis, say *xx*, will usually depend (shearing strength is an exception) on the total of the characteristics of all typical transverse planes (such as *abcd*) parallel to each other which are cut by that axis as it passes through the cube from one face to the other. The total may be (1) an arithmetical sum of the characteristics of each of the cut planes, if the particular property characteristics add (2) an algebraic or statistical sum if the property characteristics from plane to plane interact with each other (3) a limiting value if a plane with limiting characteristics sets a maximum which cannot be exceeded. By analysing the types, numbers, distributions and typical orientations of the bonds on each of a representative number of such planes each of the potential properties on the appropriate axis may be estimated. Having performed this exercise for each axis a reasonably complete picture of the overall potential properties may be built up. If some divergence appears between the properties it may be desirable to use a number of alternative possible orientations of the unit cube to determine in which direction the respective true maximum and minimum property characteristics lie, bearing in mind that the maxima and minima of all the properties in one material do not, necessarily, lie in the same directions.

Although it is easy to express in words the principles of such estimates, it is not easy to make even a very simple estimation without having a great amount of data and the use of a suitable computer. When it comes to the more complex estimates only an expert team suitably equipped can make a reasonable attempt within a useful time limit. All such estimates are based on the ideal condition of a material with the optimum possible internal arrangement of its structure and no defects, although an arbitrary correction factor may be adopted in some cases and used to give some idea of the possible effect of distributed internal defects.

Effects due to surface condition and relative proportion of surface area to volume for a given specimen of material have to be separately assessed and, in the present state of the art, usually on a more or less empirical basis.

5.8 IDEAL AND REAL MECHANICAL PROPERTIES

At first sight it would appear that the mechanical properties of a material are the properties most amenable to prediction from theoretical concepts, since such properties are the outcome of the operation of the simpler, best understood aspects of interatomic bonding and are related to reasonably predictable distributions and orientations of atoms. Within limits this is true and the ideal mechanical properties of a particular material can be postulated with fair accuracy.

Thus if we consider a typical unit cube of known orientation the reactions of the unit cube can be analysed with respect to each of the basic types of mechanical forces as if applied to each axis in turn (or simultaneously as required) and related to the distribution of the atoms and/or molecules within and between the relevant transverse planes (or longitudinal planes in the case of shear stress). Such an analysis is obviously easier and more accurate to make when there is a uniformly ordered distribution of atoms within the unit cube, as in a crystalline structure. In this case the force can be easily related to the orientations of the crystallographic planes. In the case of random arrangements, a statistical analysis relating the orientations and typical natures of the interaction forces between the atoms or molecules must be made to determine a likely representative distribution pattern of atoms that can be used for estimating the properties.

Each atom in each crystallographic (or representative plane) will exert forces of attraction and repulsion on its pattern of neighbours as any applied force moves it away from its neighbours or nearer to them (at rest there is zero average force acting between the atoms because their energies are in balance). It is possible to estimate the characteristics of these reaction forces for a particular situation and hence to estimate the likely total characteristics of the reaction forces in the unit cube. The reaction between any pair of planes will be the sum of the reactions of the atoms in one of them

$$\sigma = nf$$

where σ = the total force between the planes, n = number of atoms in one plane and f = force exerted by each atom. If more than one kind of simultaneous force is present in each plane this becomes

$$\sigma = n_1 f_1 + n_2 f_2 + \ldots n_x f_x$$

$n_x f_x \ldots$ being the numbers and associated forces of each kind of atom.

Two kinds of atom movement can be critical in these conditions

(1) pulling apart of atoms (cleavage)

(2) sliding atoms across each other (shearing)

Both movements lead to limiting maximum resistances which if exceeded give failure, the first by simple abrupt separation and the second by a more progressive jerky sliding apart. The first is possible in every material structure but the second becomes more difficult as structural order decreases because, in an unordered structure, shearing movement must be irregular and must cause some atoms to move closer to each other with a consequent rapid increase in the forces of repulsion between them. (Repulsive forces between atoms increase progressively and rapidly towards very high values as centre distance between atoms decreases below the normal stable spacing).

The average pattern of force reaction exerted on an atom in a plane subject to cleavage, tends to be similar for all situations although both its maximum value and its rate of development may differ widely with the type of atom and its situation. A typical pattern form of force on one atom is shown in Figure 5.14a.

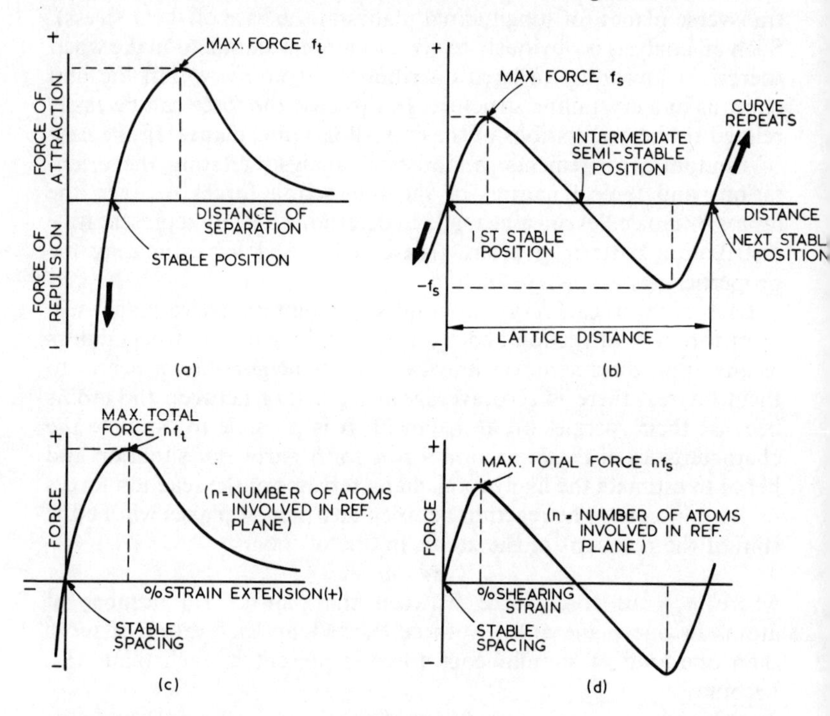

Figure 5.14. (a) Force acting on a single atom as it is pulled or pushed away from its adjacent matching plane of atoms in a solid. (b) Force on each atom in a plane as the plane is forced to slide past the next plane in a crystalline solid. (c) Total force against strain as one plane of atoms is pulled or pushed towards another plane. (d) Total force against local strain as one plane of atoms in a solid is forced to slide across another

For shearing conditions a corresponding typical force pattern is found *only in ordered structures* when it will tend to have a sinusoidal form (Figure 5.14b). This curve represents the forces between the successive positions, related to the mating lattice plane in the system, into which a particular atom could settle and give stability, provided that the whole layer, in which the atom itself lies, is able to move with it. (If an atom moved alone it would immediately meet strong repulsion from its next neighbours in the same plane.)

Each of the curves in Figure 5.14a and b shows a limiting value on the strength of the bond related to the nature of the particular type of atom and to the applied force causing displacement. If the appropriate numbers of atoms in the planes of a given system and the natures of their displacement force curves are known, the relevant ideal strengths may be calculated. The curves representing these strengths in a simple one-type atom structure will conform to the same shapes as the individual atom curves but will represent strength magnitudes corresponding to the sum of the numbers of atoms in the limiting planes (Figure 5.14c and d). Total force curve shapes for complex multi-atom systems may differ appreciably from those of the individual atoms and may not be so easy to compute.

Taking typical final force factors into account several things are apparent

(a) in a completely symmetrical crystalline system the ideal cleavage strength is significantly greater than the idcal shearing strength

(b) in a partially symmetrical crystalline or randomly arranged atomic system ideal shearing strength is likely to be higher than ideal cleavage strength

(c) plastic deformation is an unlikely phenomenon in any ideal material because with cleavage-separation failure plastic deformation is impossible and with shearing failure the energy released by elastic recovery of the system at the point where the maximum force from a particular set of stable positions is overcome, plus any energy of attraction that may be exerted by the next possible stable position (if such is available) must build up sufficient kinetic energy in the system to make the atom layer overshoot the next stable position and so on with successive positions to complete failure. (The only way to stop such a progression would be to devise a method of release of the external forces by some miraculous system that reduced the intensity of the applied force in step with the decrease in total atomic force, after the peak point in Figure 5.14d was reached, down to almost the zero value at the appropriate intermediate semi-stable position, thereafter relying on the attractive energy of the next stable position to carry the movement forward to complete an atom space step. This would have to be followed by a repetition of the cycle of application and release of stress and so on.)

It is known that in real materials the ideal strengths are not achieved except in very rare situations and that in many materials plastic deformation is readily caused.

There is another aspect of the problems found in considering the elastic behaviour of an ideal material. If a force of the relevant type, within the range of the maximum of the appropriate curve represented in Figure 5.14c or d is applied to an ideal material and then released the system will tend to spring back to its original state (elastically). If the force is released very suddenly there may be violent oscillations of the atoms about their initial positions as the elastic strain energy is released, but they will soon settle into their initial average positions. This elastic behaviour does not conform to Hook's law which states that elastic strain is directly proportional to the stress applied, but will have an elastic stress–strain relationship with a changing shape similar to that shown in Figure 5.15. Only in

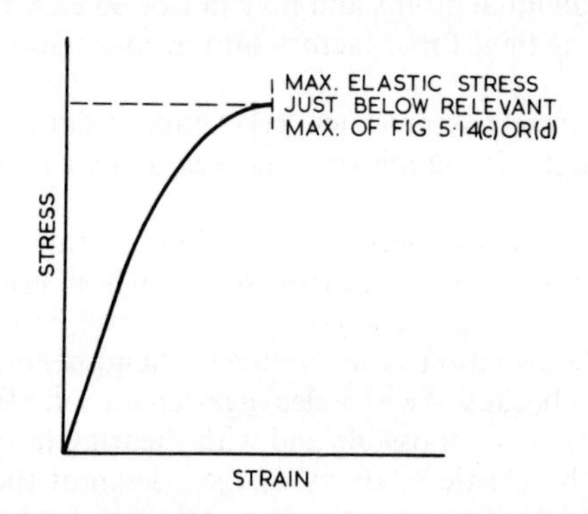

Figure 5.15. Basic form of ideal elastic stress—strain curve

compression-type loading would Hooke's law appear to be obeyed. However, in most types of real materials obedience to Hooke's law in tension and shearing is one of the most characteristic and constant features. So much is this feature characteristic that appropriate moduli of elasticity are commonly quoted and used for the study and application of each of a very wide range of materials (cf. Figure 4.10).

Can these mechanical property differences be explained in terms of the ideal? Three possible explanations might be brought forward, stress-concentration, interaction of types of force and grain boundary phenomena.

5.9 STRESS CONCENTRATIONS IN IDEAL MATERIALS

It is impossible to apply a force to a material, even under ideal conditions, in such a way that no stress concentration is caused. There are three possible reasons for this (1) local variations in the means used to transmit the force (2) unavoidable misalignment between the orientation of the structure of the material and the direction of application of the force (3) spread of the stress reaction within the material by elastic deformation of the structure about the zone of application. There is therefore bound to be some variation of reaction stress intensity within a material from the simple mean value as derived by equating external force to the area of reaction within the material. For example in Figure 5.16 the mere attachment of the loading planes at the end faces of a specimen causes some restraint

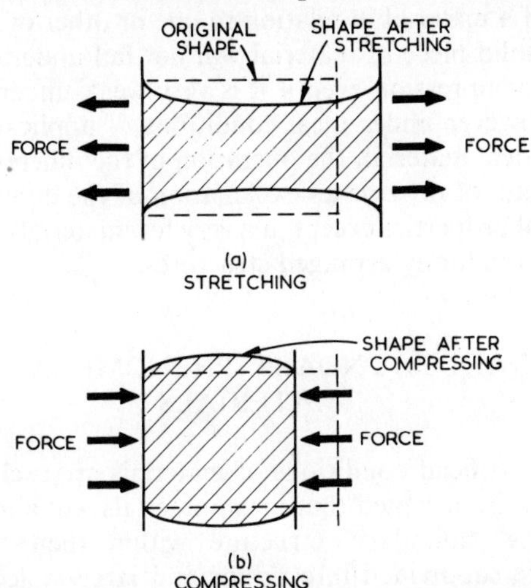

Figure 5.16. Effects of end restraint on elastic deformation

to the freedom of elastic movement of the specimen at each corner therefore a higher intensity of stress is developed in the zones near these corners. Even if the loading planes are of the same area as the planes taking the load on the specimen their material is almost certain to be different to that of the material of the specimen, so there will not be complete uniformity of action and reaction between them.

However unless stress concentrations are deliberately accentuated by influences other than these then intensities are not likely to rise much above about 2:1. This value is certainly insufficient to explain the very much greater differences between ideal and real conditions.

5.10 INTERACTION OF TYPES OF FORCE IN IDEAL MATERIALS

It was shown in Chapter 4 that no one type of force can be applied in isolation. Thus tension and compression forces each have a shearing component and a shearing force has tensile and compression components. Consequently it is possible that the complementary component forces acting (as they must) on different sets of planes to those of a given principal force could induce variations and weaknesses in strength behaviour related to the application of a principal force.

Such effects can be studied by deliberate superimposition of one type of applied force on another and can cause marked changes of behaviour of a material in relation to one or other of the forces. For example, a solid piece of material will not fail under extremely high hydrostatic compression even if it is very weak under uniaxial compression. However under ideal conditions of application of a basic force to an ideal material, the operation of the inherent complementary forces cannot provide an explanation of the differences between ideal and real properties except in a very few materials with peculiarly arranged or randomly arranged structures.

5.11 GRAIN BOUNDARY PHENOMENA IN IDEAL MATERIALS

If extremely artificial conditions of materials are excluded from the ideal it has to be accepted that ideal materials will almost invariably incorporate a 'boundary' structure within themselves. Ordered structures are subdivided into grains and large-molecule structures are influenced by the surface contact relationships between the molecules. Boundary behaviour and orientation differences between adjacent zones of the base material will greatly influence both stress concentration effects and type and behaviour of stress reaction from an applied force, increasing the intensity of stress concentrations and increasing the possibility of initiating failure in a weak zone. Nevertheless these effects alone can only rarely explain the divergencies between ideal and real properties and then only in materials with excessively wide divergencies between the mechanical properties in different parts of their structure.

There remains the natures of the boundaries themselves and this would seem to give the required answer. After all, if grain boundaries give way at strengths far below the ideal strengths of the main body of material this could explain low strength phenomena. Even plasticity might be explicable if the relatively unstable narrow zones around grain boundaries behaved like liquids and allowed grains to slide past each other. Grain boundary weakness was once accepted as the whole explanation, until it was discovered that except in a very few cases grain boundary zones do not flow like a liquid and that except in cases of extreme boundary weakness boundaries are not usually the main places of failure in a real material.

It is unquestionably true that a grain boundary zone is usually weaker than the main body of a material. This is to be expected from the three facts that (a) bond density must be lower in a boundary zone than in the body of a material (b) individual bond strength must be lowered because atomic positioning is not as regular or positionally stabilised as it should be and (c) low strength impurities tend to concentrate in grain boundaries. Offsetting these weaknesses is the geometry of boundaries which tends to minimise some of their effects.

Grain boundaries are rarely uniform in distribution or direction (Figure 5.17) therefore if failure starts to propagate along a boundary under an applied stress the failure cannot travel far in any one direction before meeting a marked change of direction which is likely to arrest its progress. Similarly if a boundary permitted sliding to take place, the sliding could not progress more than a very short

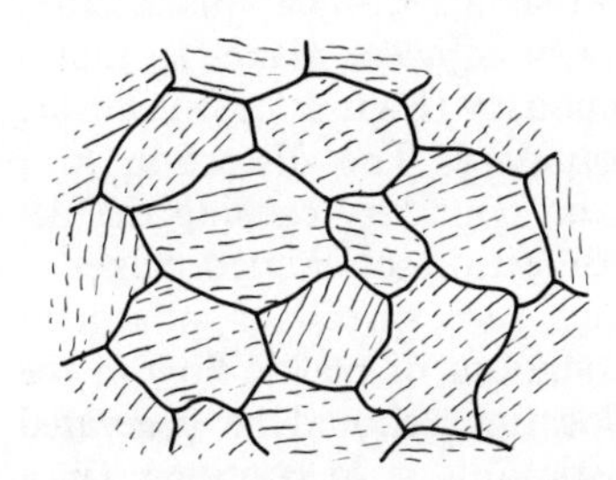

Figure 5.17. Irregular distribution of grain boundaries

distance before it would be arrested by the need for grain or molecular shape to be changed to fit the movement.

No doubt in an ideal material its grain boundaries would be its centres of failure, but their weaknesses are still insufficient to explain the very much lower strengths of real materials.

5.12 MECHANICAL PROPERTIES AND DEFECTS IN REAL MATERIALS

The true explanation of the limited and modified mechanical properties of real materials usually rests entirely on the presence and/or development of numbers of structural defects. In materials with ordered structures the defects are dislocations. A dislocation is relatively easy to move. If an end view of an edge-type dislocation is as shown in Figure 5.18a, it is easy to see that quite low stress

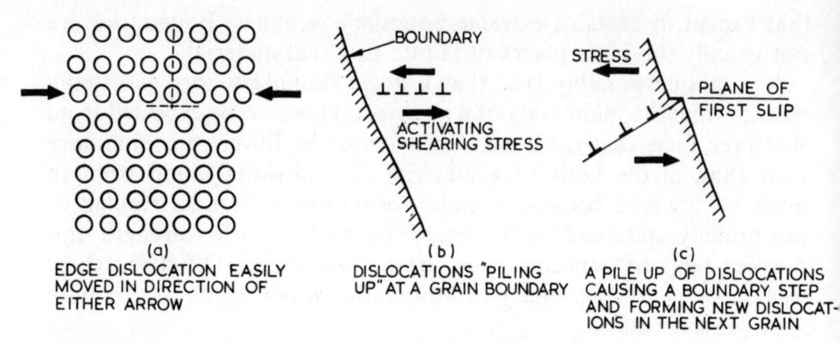

Figure 5.18. Interaction at grain boundaries of stress activated dislocations

of the appropriate kind will cause the defect to move along the plane until it reaches the boundary. At the low activating stress it is unlikely to pass through the boundary (unless the latter is an outside surface) but will be held there as long as sufficient stress is applied. The same stress may cause a number of similar dislocations to move up either in the same plane or in closely adjacent planes so that a dislocation pile-up develops at the boundary (Figure 5.18b) causing accumulated distortion near the boundary. The distortion may become sufficient to break through the boundary causing the development of similar dislocations on differently orientated planes in the adjacent grain (Figure 5.18c). Each such spread of movement that occurs causes only a very small amount of plastic flow in the material and it is essential that dislocations should be generated spontaneously in the structure if plastic flow is to continue, since the number of dislocations present in the structure, although numerically large, is always insufficient to give more than a minimal amount of total plastic distortion. However generating mechanisms or sources (sometimes called *dislocation mills*) are fairly common.

A potential source can be illustrated from Figure 5.2c; if a dislocated area is imagined as pinned in a crystallographic plane which is intersecting a crystallographic plane lying in the plane of the paper, the intersection of that dislocation being represented by the dashed line AB (points A and B representing the intersection points of the screws bounding the pinned dislocation). In this situation quite a small shearing stress applied to the plane lying in the plane of the paper could start a dislocation moving between AB in the same plane. With a sufficiently high stress the dislocation could burst out from between AB (see stages (2) and (3) of the diagram) spontaneously forming both a closed loop that will move to the plane boundaries and another dislocation that will complete the movement back to the line AB, which is then available for the generation of another dislocation if stress intensity is maintained. Dislocations will continue to generate at AB as long as a sufficiently large shearing stress intensity is maintained against the increasing back-stress from the generated dislocations as they react with the nearest boundary, until the mechanism is used up (each dislocation must climb one step further up A and B as it swings around them). Several kinds of such source mechanisms are possible, the intensity of shearing stress required to operate them varying inversely with the length (AB) over which the dislocations form and being related also both to the characteristics of the particular lattice and to the total area of the plane in which they can move (the larger the plane the slower the build up of back stress).

The combination of the low strengths, limited elasticities, relative uniformity of the respective moduli of elasticity and the marked plasticities of many crystalline materials, notably metallic materials, lies in the operation of dislocations. When stress is applied to such a material a significant amount of dislocation movement begins at a particular stress intensity. The yield stress is very much lower in intensity than the peak stress indicated on the ideal stress–strain curve. This yielding starts so much plastic flow within the material that the ideal elastic behaviour cannot be developed (Figure 5.19).

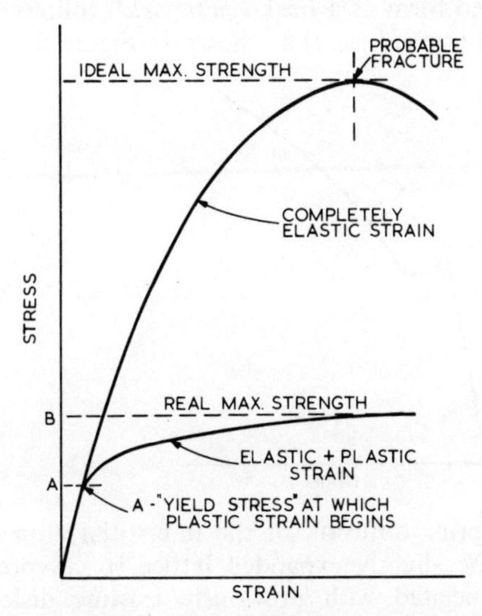

Figure 5.19. Comparison between ideal and real mechanical properties

Thus the full elastic range of the material is limited, by plastic flow, to stress level *A* on the relatively straightest part of the curve and the real strength B is a compromise between the ease of continued flow of dislocations and the stage at which critical failure-causing stress concentrations are set up.

If critical stress concentrations build up quickly failure begins at a low applied stress and may reach completion so quickly that very little plastic strain is developed. This tends to happen in materials with structures in which dislocation movement is greatly restricted by such things as the presence of rigid or brittle constituents or impurities or by the presence of such severe distortion in the structure that dislocation movement requires very high stresses.

In a material in which continued dislocation movement is possible, the stress required to cause continued plastic flow will rise progressively with flow as the various avalanches of dislocations tangle with each other and with other obstructions such as boundaries, and as the lower energy sources get used up (Figure 4.9). This progressive rise in flow stress is known as *strain hardening* or *work hardening*.

Materials with ordered structures, containing relatively small-sized interstitial solute atoms and capable of manifesting plastic flow, are often able to exhibit another strain phenomenon known as a *yield point* or *stepped yield*. This effect appears on the stress–strain diagram either as a simple step on the slope (Figure 4.10) or in its fully developed form as a peak (*first yield*) followed by a drop in stress to a flat (*second yield*) as shown in Figure 5.20. This result is

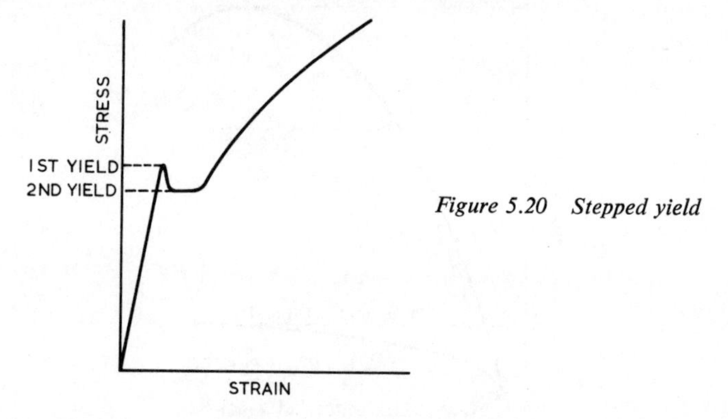

Figure 5.20 Stepped yield

produced by prior diffusion of the interstitial atoms, through the lattice, into the slightly expanded lattice spaces produced by the distortion associated with previously existing dislocations. Such atom migration, said to cause a *foreign atom atmosphere* about dislocations, stabilises the affected dislocations making them more difficult to move. When sufficient stress is applied to cause the dislocations to move the solute atoms cannot follow quickly enough to keep up, so the stress needed to keep them moving drops and dislocation movement can continue at the lower stress until the normal processes of interaction begin to catch up and strain hardening begins. This effect should not be confused with a similar yield effect which appears when plastic flow first begins under stress applied to a single crystal located in such a way that slip is directed, initially without cross slip, into a particular set of planes of *easy glide*.

Materials with rigid network structures or ordered structures which do not permit dislocation movement (e.g. the diamond structure of carbon) are much more sensitive to other defects such as vacancies and fissures. This sensitivity occurs because plastic flow is not available to relieve stress concentration and so very severe stress concentrations can develop in the vicinity of defects. Brittle failure at low nominal stresses readily occurs in such materials, particularly if shock loading is applied.

Those materials which do nearly develop their ideal strengths are usually the ones with low strength-expectancy associated with weakly-bonded internal boundaries.

5.13 IDEAL AND REAL ELECTRICAL PROPERTIES

The electrical properties outlined in Chapter 4 are directly related to the interactions between the valency electrons of the different types of atoms. Therefore if the types of bonds within the material are known, the contribution each one can make to the electrical properties is known. It is feasible to estimate for an ideal material the numbers of suitable bonds that should exist per unit volume, and the orientations of those bonds and hence the ideal electrical properties. Materials with metallic bonds will tend to have low resistivity and materials with predominantly ionic or covalent bonding will tend to have high resistivity. The principal factors are the potential mobility of electrons within the material and the minimum energy required to cause electrons to jump the energy barriers that hinder interatomic electron movement.

Anything which tends to disturb the uniformity of electron movement will make interatomic movement less easy. For example, increased thermal energy, in so far as it causes more turbulent movement of electrons, increases resistivity. In a like manner, most structural defects increase resistivity by disturbing and/or interrupting the patterns of electron movement. There are exceptions to this for example, when the defects in the structure are impurity atoms which are themselves more conductive than the parent atom or which can provide electron bridges between the parent atom orbitals by lowering the energy barriers between them.

When electrical conductivity in a material is mainly one of ionic transfer, the presence of vacancies and/or dislocations can provide easier diffusion paths for the relatively large ions to travel through and so reduce resistivity.

The general effect is that the presence of defects usually increases the resistivity, particularly of low resistivity materials, and in some cases reduces the resistivities of high resistivity materials.

5.14 IDEAL AND REAL MAGNETIC PROPERTIES

Magnetic properties are probably less predictable than any of the other physical properties, since less is predictable about the ordering of magnetic domains, but it is likely that impurities will upset the flux distribution in the local fields within domains. In general, defects will make overall magnetisation less easy to achieve and they may also make the overall balance of magnetisation in an unmagnetised material less uniform so leaving a slight polarisation of the total magnetic field.

On the other hand, if the individual magnetic domains in a normally soft or weakly magnetic material can be ordered on one axis and the defects in the material so distributed that the domains are prevented from reordering themselves, then the material might be converted into a strongly-magnetic hard magnet. This can sometimes be achieved in certain materials by producing a geometric distribution of suitable grain boundaries that gives both a very fine grain structure and an appropriate shape and orientation to the grains. In other materials a similar effect can be achieved by embedding ordered magnetic particles, of about magnetic domain size, in a rigid matrix so that the magnetic fields of the particles are cumulative.

5.15 IDEAL AND REAL THERMAL PROPERTIES

In parallel with the differences between ideal and real electrical properties one would expect thermal properties to be similarly affected by defects. Indeed thermal conductivity is almost always reduced at all levels by the presence of defects except in cases where electron transfer between atoms is made easier by the presence of foreign bridging atoms, or where the mechanical rigidity of a structure is so increased by extra bonding of foreign atoms or molecules that elastic vibrational transfer of heat energy becomes easier. The latter effect is not usually very marked in its intensity unless the increased number of bonds is large.

The ideal coefficient of thermal expansion may also be influenced by defects. If the presence of a defect reduces the density of a material then it is to be expected that the coefficient of thermal expansion and contraction of the material would tend to decrease since the change of density implies fewer atoms to interact in a given volume. On the other hand if defects such as dissolved impurity atoms give a greater density to a material by filling gaps in the structure it is likely that the coefficient will be raised. In neither case is the change either of density or coefficient likely to be very marked.

Certain defects including some types of impurity atoms and molecules and particular configurations of dislocation defects in crystalline materials can greatly influence thermally induced melting and solidification phenomena, particularly by affecting nucleation during solidification. The latter influence makes solidification both easier to begin and faster to complete by a two-fold effect. The first effect is that the presence of a suitable foreign atom or molecule can disturb the local energy balance between the parent atoms and/or molecules in a cooling liquid in such a way that it makes it easier for a relatively stable group of the parent units to accumulate in its vicinity and so form a nucleus. This process is known as *heterogeneous nucleation.* The grouping of parent atoms in such a nucleus is not as orderly as in *homogeneous nucleation* (when no foreign agency is present) and this irregularity causes the second effect by making it easier for other atoms or molecules to add themselves to the nucleus so increasing the rate at which solidification growth will proceed. A relatively small number of suitable impurity atoms is all that is required to give a greatly increased rate of nucleus formation in a cooling liquid at the beginning of solidification and a consequent increase in the number of independent grains found in the completely solidified material. This process has two complementary results (1) because grain is small the ratio of surface area to volume of grains is raised and there is a higher proportion of grain boundaries in the solid materials (2) defects are grown into the grains with the disorder caused by the incorporation of the foreign inocculant. In a crystalline material such defects are mainly in the form of dislocations.

The most obvious effect of heterogeneous nucleation is to reduce the amount or suppress the the phenomenon known as *undercooling*, which results from the normal difficulties of homogeneous nucleation. Because homogeneous nucleation is difficult to start in a pure, slowly cooling, elemental liquid (as shown by the low nucleation rate indicated for just below the melting temperature in

Figure 5.7b) the liquid will undercool; its temperature will drop below its theoretical solidification temperature before solidification begins. Subsequently its temperature is likely to rise transiently towards the melting temperature as thermal energy is released from the stabilising mass of atoms, giving a phenomenon known as *recalescence*. If the rate of nucleation is increased by heterogeneous nucleation by introducing trace quantities of a suitable innoculant the amount of undercooling decreases and, if the heterogeneous nucleation is made very easy, it may disappear altogether leaving only a step on the cooling curve at the melting temperature level, as shown in Figure 5.21b. Associated with nucleus formation is the

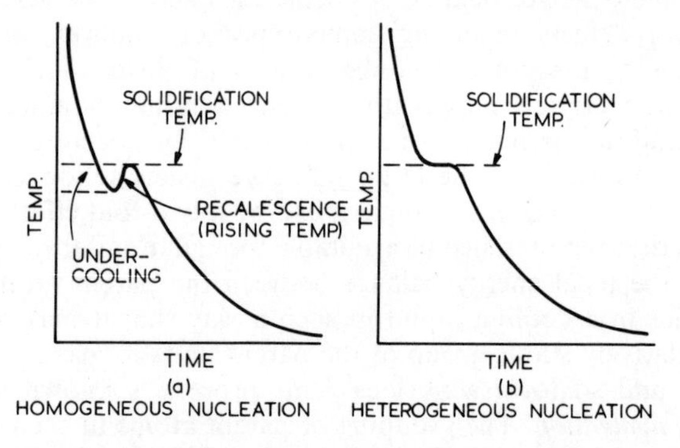

Figure 5.21. Effect of nucleation mode on solidification

effect of rate of growth from a nucleus. Because of the orderliness of its atom stacking, an ideal nucleus will not grow rapidly since there are not many energy gaps or 'holes' on its surface into which cooling atoms (or molecules) can settle. Therefore its solidification rate is very low even after nuclei have formed. However in heterogeneous nucleation the nuclei are not so orderly and there are more energy gaps on their surfaces into which additional cooling atoms can settle. Since the filling of energy gaps cannot be perfect and will leave new gaps on the expanding surface, as layers are added, solidification growth may continue rapidly and progressively, once it begins, until all available atoms are taken up. Once solidification begins, either homogeneously or heterogeneously, the temperature of the mass will tend to stay constant until solidification is complete, unless recalescence is strong.

Melting of a solid is also a nucleation process and local melting will not begin properly until a sufficient number of mobile atoms on an exposed surface have coalesced into a surface droplet large enough to begin to spread by absorbing atoms or molecules from the surface of the solid (Figure 5.22). However this process is not nearly so difficult to initiate homogeneously as the solidification process so *superheating* during melting is not usually as great as undercooling during solidification and the associated *decalescence* (transient fall

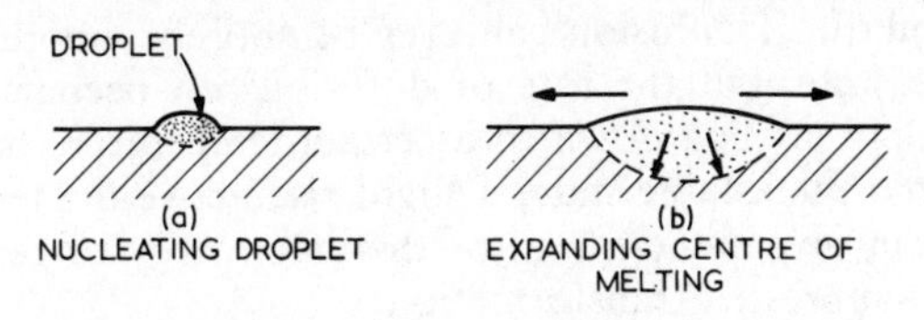

Figure 5.22. Melting

of temperature back to the melting temperature as a solid melts) is correspondingly less than recalescence. Nevertheless heterogeneous nucleation does speed the melting process.

Similar effects are observed in complex materials containing more than one kind of atom and/or molecule, but the effects may be modified by such materials having a temperature range of melting or solidification instead of a fixed melting and solidifying temperature.

Another factor is the effect of rate of heating and cooling. Slow to slightly faster cooling or heating through a change-of-state temperature does not usually cause any increase in either the rate of nucleation or in the rate of progress of the change after nucleation has begun. But more rapid change exaggerates undercooling and superheating differences although the former is always more influenced than the latter.

Structural changes in the solid state take place by similar phenomena of nucleation and growth. This kind of change begins at nuclei which nearly always form on or near a grain boundary or interface where there is disturbance to structural uniformity. There is delay in growth from these nuclei, governed mainly by the limiting rate or rates of diffusion either of self-diffusion of parent atoms or of interdiffusion of reacting constituents whichever is relevant. Several kinds of defect can speed such movements, in particular circumstances, by making diffusion more possible but rapid temperature change may still be able to swamp their effects, if the level of activating energy required to operate them is significant. Rapid temperature change can create internal structural defects such as interstitial and substitutional foreign atoms if such atoms are retained in the structure in a condition of *supersaturation*, after a temperature-change-induced reduction in solubility, because the diffusion energy required to move them away in the new state of the material is greater than the residual strain energy in the structure which is tending to cause the movement. *Metastability*, as this situation is called, develops most commonly with falling temperature, because then the rigidity of a structure is likely to be increasing, making diffusion that much more difficult.

It is doubtful if diffusion can ever be entirely suppressed in any metastable state, but the rate of diffusion can become very slow. Often if rapid cooling down to a certain temperature has caught a material in a metastable state, a slight rise above the terminal temperature can give sufficient extra thermal energy to reactivate the seemingly suppressed transformation.

5.16 IDEAL AND REAL CHEMICAL PROPERTIES

Although chemical properties cannot be specified as simply and definitely as physical properties (see Chapter 4) they may be greatly influenced by defects in the material under consideration. Any chemical reaction whether it is between solids or between a solid and a liquid requires diffusion to take place in the vicinity of the reaction interface if the reaction is to be progressive and not die. This diffusion barrier may be overcome, or lowered, artificially in deliberately stimulated chemical reactions by using such means as (1) preparing the reactants in a suitable form such as a liquid or powder (2) by intimate intermingling of the reactants (3) by mechanical agitation during reaction.

Most kinds of defects likely to be present in a real material can influence diffusion in one way or another and hence can influence chemical properties either by changing the rate of reaction or by changing its mode of progression (e.g. a general interfacial reaction may be changed to a localised pitting reaction in which the reaction boundary spreads preferentially in certain localised areas into one or other of the reacting substances (Figure 5.23). Defects such as

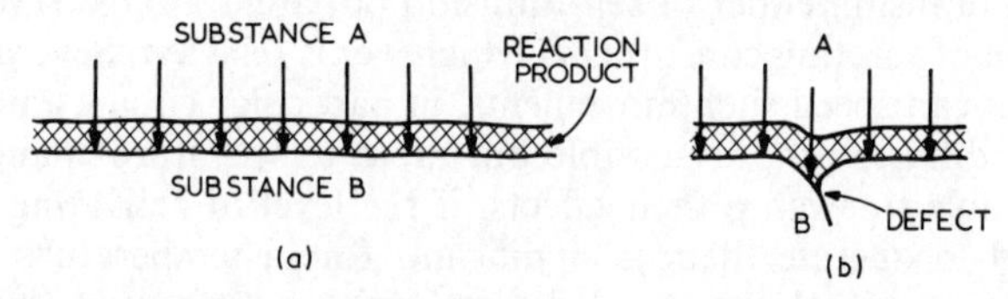

Figure 5.23. (a) General progress of a reaction at an interface, A diffusing through product to B. (b) A diffusing faster into B along a defect giving a pit

vacancies and, more particularly closely spaced vacancies, can aid diffusion of atoms, or even molecules, of a relatively large size in appropriate conditions. Phase boundaries usually provide easy diffusion paths in any substance in which they are present. Free surfaces at internal boundaries, or cavities, may also provide good diffusion paths. Dislocations provide potential diffusion paths for relatively small atoms. The possibility of increased rates of diffusion may not always lead to a speed up in a reaction particularly if the atoms that diffuse most readily are not ones that contribute to the reaction. In some cases diffusion of neutral atoms may even oppose a reaction if they are carried into the reaction zone and there act as inhibitors. Whatever the situation it can be seen that the ideal chemical properties are likely to be changed by the real nature of materials.

However there are likely to be other influences affecting the progress of a reaction and so tending to modify ideal behaviour. Among such influences are (a) the increasing proportions of the reaction product (b) the influence of the reaction on environmental conditions (c) the influence of other constituents present in the environment. Each of these aspects is outlined briefly in the remaining parts of this chapter, but no categoric statement can be made about the effect of any one of them compared either with the others in the group or with the effect of defects. Each case has to be considered on its own merits.

5.17 THE INFLUENCE OF REACTION PRODUCTS ON CHEMICAL REACTION

In Chapter 4 mention is made of the effects of the build-up of the product of a reaction between the surfaces of the substances from which the reactants are being derived. These effects are fairly obvious but there is another aspect of the same mechanism, notably when two phases react to produce a third phase and in doing so release one or more other phases. For example, in the deoxidation of a molten steel during refining, manganese may take part in the following reaction

FeO (an undesirable oxide of iron) + Mn = Fe + MnO

The manganese has taken up the oxygen from the oxide of iron to form an oxide of manganese and leave the iron free. In this case it is the by-product of the reaction, the iron that is required and, fortunately, it is possible to float the manganese oxide clear of the liquid metal into the slag (the non metallic residue left floating on the surface of a liquid metal after a refining operation) and so remove it. But what happens if an unwanted product of a reaction does not remove itself by diffusing away and cannot be removed by any convenient external aid? In this case the surplus product may accumulate in the reaction zone and, according to its nature, do one or more of several things.

(1) It may decrease the rate of interdiffusion of reactants through the reaction zone and thus damp down the rate of reaction, perhaps bringing it to a stop.
(2) It may react with another residue of the reaction or another, previously unaffected phase, present in the vicinity producing still another by-product which in its turn may hamper the original reaction or cause the development of the third eventuality.
(3) It may, by its own inherent weakness or by its interference with the bonding in the reaction zone, produce physical weakness in the latter, or even disintegration, if such changes are not already occurring for other reasons.
(4) It may either, by its presence unbonded within the reaction zone or by diffusing away to leave microscale porosity, provide better paths for diffusion through the reaction zone and so speed up the rate of reaction.
(5) It may have little or no effect on the rate of reaction but may remain entrapped in the desired reaction product and have adverse effects on the subsequent use of the latter, perhaps giving rise to corrosive disintegration.

5.18 THE INFLUENCE OF A REACTION ON ITS OWN ENVIRONMENT

All chemical reactions between materials cause changes in (1) the energy balance between the constituent atoms (2) the relative total volume of the participating atoms (3) the physical properties of the substances involved. Such changes can cause changes in the environmental conditions, some changes being transitory and others permanent in nature.

Exchange of thermal energy causes transitory temperature changes. Thus if thermal energy is released (*exothermic reaction*), temperature will rise locally until the surplus heat is dispelled. If thermal energy is absorbed (*endothermic reaction*) the temperature will fall. Since the ambient temperature can affect the rate of reaction, then if the amount of temperature change is large, the reaction itself will induce a change in its own rate. Most commonly a rise in temperature accelerates a reaction and a fall retards it. The actual effect may be the result of a direct change in the rate of reaction but it could also be caused by a change in rate of diffusion of the reactants (rate of diffusion often varies with change in temperature).

Change of total volume during a reaction in a structural material is likely to generate a local stress between the reaction zone and the unreacted material. The nature and distribution of the stress will depend on the magnitude of the volume change, its nature (expansion or contraction) and the mechanical properties of the phases concerned. A large expansion in a reaction product could lead to the build-up of heavy compression stresses and possibly result in cracking in the surrounding material which would be stressed in tension. A large contraction in a reaction product might induce sufficient tensile force within the product to lead to its internal fracture under the tension. Such stresses can directly change the rate or even the nature of a reaction. The extreme case is found in the detonation of an explosive material when the self-generated compression wave of the reaction associated with the initial chemical change in the material accelerates the rate of spread of reaction at an extremely high rate.

Release of a gas during a reaction may also modify the environment of a reaction by displacing an atmosphere which may initially have been contributing to the reaction. A released gas may also accumulate in defects, building up pressure, locally modifying the rate of reaction and initiating fracture. One gas which is very liable to behave in this manner is hydrogen which in its atomic form (H) diffuses readily through many structures and accumulates in any defect space within the structure, there it combines to the molecular form (H_2) and builds up pressure, sometimes to a level at which cracks may propagate.

The aspect of changed physical properties in the vicinity of a reaction is of interest here only in as far as it affects environment. It can be appreciated that a reacting product which fractures as it develops can have a marked effect both on the diffusion mode within the zone and possibly also on the direction and mode of spread of the reaction.

5.19 INDIRECTLY INVOLVED CONSTITUENTS AND CHEMICAL REACTIONS

Chemical reactions within a material or between two materials very often involve constituents which are incidentally present. This is noticeably the case when only one phase of a multiphase material is reacting with another or when impurities are present. Such neutral constituents have one or more of a number of possible effects, the relative incidence of these depending on such features as (a) the atomic nature of the extra constituent (b) the geometry of the distribution of the extra constituent relative to the reactants (c) the ambient conditions in the vicinity.

There are four possible principal effects of an extra constituent grouping into two pairs, one pair associated with decreasing rate of reaction and the second pair with increasing rate of reaction. Any effect may operate alone or in conjunction with one or more of the other effects, the final overall result being an average of the active individual effects. The four principal effects are as follows.

(1) An extra constituent may slow down a reaction by a simple dilution of the reactants which reduces the average density and also increases the average distance over which reactants must diffuse to meet each other.

(2) An extra constituent may slow down a reaction by blocking existing diffusion paths or by sharing these paths with the reactants and so reducing the potential average frequency of travel of reactant(s) along them.

(3) Suitable extra constituents may increase the rate of a reaction by increasing the potential rates of diffusion of reactant(s) if the extra constituent opens a diffusion path for the reactants easier than those paths already available.

(4) A suitable extra constituent may act as a catalyst and speed up a reaction. (Catalysis is not simple diffusion, therefore it is separated from effect (3).)

Each of these effects may be harmful, or helpful, in differing circumstances. An effect can be very harmful if it operates unintentionally but it may be very beneficial if it is used deliberately. Impurities in a material and unexpected constituents in a service environment are frequent causes of harmful effects.

BIBLIOGRAPHY

MARTIN, J. W., *Elementary Science of Metals*, Chaps 4–6, Wykeham, London

TRELOAR, L. R. G., *Introduction to Polymer Science*, Wykeham, London

TWEEDDALE, J. G., *Metallurgical Principles for Engineers*, Chap. 3, Iliffe, London (1962)

VAN VLACK, L. H., *Elements of Materials Science*, Chaps 4–8, Addison-Wesley

6 Control of the Useful Properties

It is rare to find natural materials with all the requisite properties to suit even one simple application. We are then forced to consider the modification or adaptation of existing natural materials or the synthesis of new materials.

There are four ways to deal with the problem of matching a material and its properties to a desirable application

(1) Adapt or modify the microstructure of a natural material to suit a specific application.
(2) Make simultaneous use of two or more separate materials, each with some of the required properties.
(3) Use integrated materials of two or more kinds, combined in one matrix.
(4) Develop new structures by physical reaction, chemical reaction and synthesis.

In the remaining parts of this section the limits of each of these respective ways are outlined and then in subsequent successive sections each way is surveyed briefly with respect to method and results.

It must be borne in mind that these ways need not be used in isolation and indeed many materials are the result of combinations of two or more of these approaches.

One outcome of the problems of matching material to application is that commercial materials often tend to be classified into specific groups relating the particular type with the application for which it was first developed (fireclay, corrosion-resistant-steel etc.). This kind of classification is usually vague and often misleading. Materials in a seemingly specific group are often widely used for alternative applications. The governing factor, more often than not is cost of production, and a relatively unsuitable but cheap material, giving a very short service life, is often used in preference to a better material costing more. This fact explains many of the failures commonly experienced in the use of domestic utensils and private cars.

6.1 ADAPTATION AND MODIFICATION OF STRUCTURE

In this group are all those methods by which a natural material has its structure adapted in some way to enable it to develop latent properties. Most commonly this group relates to metals, which with their particular crystalline structure and notable sensitivity to the presence of defects are unable to exhibit more than a small proportion of their potential ideal properties, particularly their mechanical properties. By manipulation and adjustment of the structures of such materials the properties may be improved.

6.2 COMBINED USE OF SEPARATE MATERIALS

This way of suiting materials to purposes consists of using separate materials in a geometric distribution, within a planned construction that enables each material to exercise its most characteristic properties without being overtaxed with respect to its deficient properties. Ideally this way works best when operating conditions are also adapted, perhaps by changing the principal of design and operation of a construction, to give a more selective distribution of environmental conditions. Examples will be given later in this chapter.

6.3 USE OF INTEGRATED MATERIALS

A logical development of the combined use of separate materials is to bond suitable mixtures of such materials into a coherent whole without the individual materials losing their identities, but in such a way that the integrated materials may be treated and manufactured as one material. This system results in saving in cost and in complexity of construction and often enables the development of a reasonable and efficient compromise between the individual properties. This system is in very common use and has probably resulted in the development of more materials of construction than all the other systems together.

6.4 SYNTHESIS OF NEW STRUCTURES

To produce an entirely new material ideally suited to each new application is a very attractive idea but several factors operate against its fulfilment in practice. (1) Research and development costs are high. (2) A considerable unpredictable time interval may elapse before a reasonably effective solution is found. (3) A really effective solution may not be possible and all the costs of preliminary work will be wasted up to the time this is realised.

More commonly, as a result of general research new potential material sources are discovered, new methods of synthesis or treatment are developed and understanding of particular principles underlying material behaviour increases. This leads to a more or less random discovery of new systems of synthesis. At this stage a new type of material that looks promising for development may be adopted by a research and development body and a systematic attempt made to determine its optimum properties, its potential use and to distribute information about it to potential users. A use is more often fitted to a material than is material developed for a specific use although there are exceptions.

6.5 ADAPTATION OF THE STRUCUTRE OF A MATERIAL

Nearly every material, whether extracted from nature or synthesised, is produced initially with its ideal physical properties undeveloped, usually because of numerous inherent structural defects. The relative effects of the forms of defects depend on the nature of the materials's structure. Thus within the very closely packed orderly structure of a metal crystal, dislocation defects have a major effect, whereas in the structures of other crystalline substances their effect is diminished and, in less ordered molecular structures, negligible. On the other hand microfissures may have very marked effects on the properties of materials with rigid molecular network structures and in non metallic crystalline structures but very little effect on the properties of many metallic materials. Two things are certain: the ideal properties cannot be exceeded without changing the nature of the structure and the ideal properties cannot be approached without either removing all the defects or by controlling some of their influences. Thus adaptation resolves itself into two lines of approach: refinement of structure to eliminate defects and control of the defects within the structure. Normally, both lines of approach are used together.

6.6 CAUSES OF DEFECTS

Most defects found in materials originate in the modes of initial formation. Every such mode operates on a principal of nucleation and growth, with all its potentialities for random structural orientation within the resulting material. Difficulty in generating nuclei in a manner suitable for reasonably rapid formation of the required form, usually solid, of the material often makes it necessary to induce heterogeneous nucleation by introducing innoculants in the form of foreign atoms or molecules. These innoculants are themselves defects but they create disturbances to the uniformity of the local structure, which aid the required process. The disturbance results in the development and perpetuation of various associated microstructural defects culminating in surface boundary defects where independently nucleated volumes meet.

In addition to such inherent defects there are other defects which are introduced as a result of the industrial means used to deal with bulk production. Whenever a scale of production beyond the laboratory scale is attempted, the manufacturer comes against practical limitations to his process.

(1) Containing vessels have to be larger and it may not be possible to make them effectively or economically of the most suitably unreactive materials which would have been used in the laboratory work.
(2) Reacting substances used to separate and purify a material from the contaminants with which it is associated in its natural state now have to be used in a bulk form which is less accurately controlled and likely to be less pure.
(3) With increasing bulk it becomes less easy to control the prevailing conditions such as temperature, pressure and atmospheric composition.
(4) With increasing bulk of manufacture it becomes progressively more and more difficult to keep the place of manufacture free from contaminating dirt.

6.7 REMOVAL OF DEFECTS BY STRUCTURAL REFINEMENT

Since impurities and other structural defects are harmful to the properties of a material it is desirable that a material should be as free from them as possible. How can reasonable purity be achieved? The most obvious first step is to start with raw materials as free from impurities as possible. With synthetic materials this is relatively easy if the basic chemicals are prepared in a pure state and subsequently synthesised in an uncontaminated environment. It is not so easy with materials derived from natural sources. Such materials always contain variable amounts of impurities and the best that can be done is to use the source which either contains the lowest average proportion of contaminants or else contains the most readily removed kinds of contaminants.

Subsequently, a natural or synthesised impure raw material is prepared for further treatment by reducing it to a powdered, paste or liquid condition. In this state it can then be intimately mixed with powdered solid or liquid *flux* (a chemical agent for decontamination by reacting with and separating out harmful impurities) and other reaction agents and treated to give the appropriate chemical reaction. Flux reaction produces a *slag* which has to be removed from the purified material along with any unused flux that may be left. Sometimes slags may form useful byproducts but more often they are useless wastes, the safe disposal of which adds to the final cost of the treated material. More than one fluxing operation may be necessary to enable a flux to carry the reduction of the impurities far enough to achieve a required degree of purity. Alternatively, it may not be possible to develop one flux that will react on all the kinds of impurities in which case it may be necessary to make successive use of differing fluxes to get rid of differing impurities or even different states of the same basic impurity.

Another method of refining is to use electrolysis of prepared electrolytic, liquified, raw material (liquified possibly by melting or by dissolving in a suitable carrier). This system when practical can produce a pure raw material from a difficult and unpromising natural source (e.g. aluminium is separated from powdered bauxite—natural alumina—by electrolysis after dissolving the alumina in a bath of molten sodium aluminium fluoride).

Distillation of liquefied minerals or other source materials is another method for separating raw material out of a complex matrix, particularly if more than one useful material can be separated out in one sequence of operations. Some mineral ores contain small proportions of noble or rare metals which can form useful distillation byproducts when extracting larger quantities of more common materials by this means. Probably the best known common industrial distillate is petroleum, but certain distillation byproducts from crude oil also provide the basic raw material for making a range of synthetic materials. A notable example of this kind is ethylene which forms the basis of a number of polymeric materials.

Raw materials which are solid at n.t.p., but which can be melted by heating, may sometimes be refined further by cycles of remelting and resolidification which give entrapped slag inclusions a better opportunity to float clear or give more active fluxes a chance to mix in.

The required standard of purity varies with the application. For most constructional materials an impurity content not more than

1 part in 10^5 of the required material is considered extremely good, often being described as *super pure* but for many applications a lower standard is acceptable. On the other hand for certain specific materials, notably semiconductor materials for electronic applications, this standard is not nearly good enough and an impurity content *less* than 1 part in 10^8 may be required.

Refinement to this last standard is a highly specialised and costly business in relation to the cost per unit weight or volume of the material and can usually be effected only on small quantities of material at a time. Probably the most effective and generally used method of refining to this kind of standard is *zone refining*. This method is applicable only to materials that can be melted readily by heating, preferably electric induction heating because of its rapidity. The method relies for its effect on the fact that the solubility of a solute contaminant in a solvent base material usually changes in a controlled manner with the temperature of the liquid solvent. As a result, if a band of the solvent material is liquified and then traversed through the volume of the solvent (see Figure 6.1) it

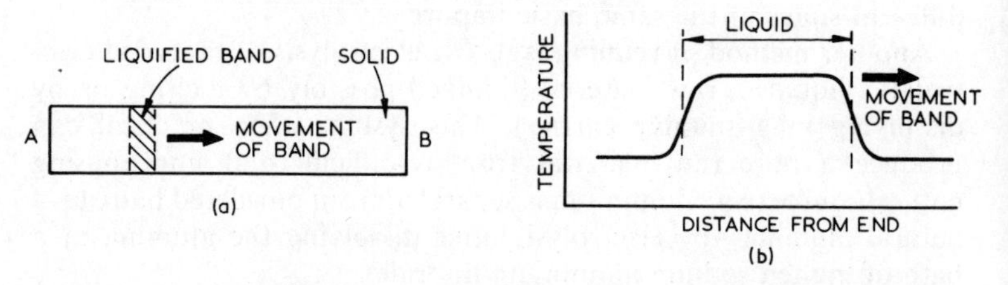

Figure 6.1. (a) Arrangement for zone refining. Solutes with decreasing solubility carried towards B and solutes with increasing solubility with falling temperature towards A. (b) Rise and fall in temperature has a pumping action on changing solutes

either carries a small quantity of the solute forward with it or drains a small amount back towards the freezing face. In either case solute impurities can be conveyed step by step by a series of liquid zone passes to one end of a uniformly shaped solid bar of the solvent material. The cost of this method is high both because it is relatively slow compared to less effective methods and because it can be applied only to small cross-sectional areas of solid material (the melting heat must be able to penetrate right through the thickness of the material to be effective). Under rather special specific conditions refining by repeated electrolysis or redistillation can produce similar degrees of purity in small quantities of some materials but the materials to which such treatments are applicable are not normally required in such a state of purity for constructional use.

For a raw material to be cheap it must be possible to extract and prepare in bulk. Hence, iron ore is treated a million kilogrammes or more at a time in a piece of equipment such as a blast furnace (Figure 6.2). It is regularly or continuously charged with powdered ore ready mixed with fluxing agents (such as limestone) and a

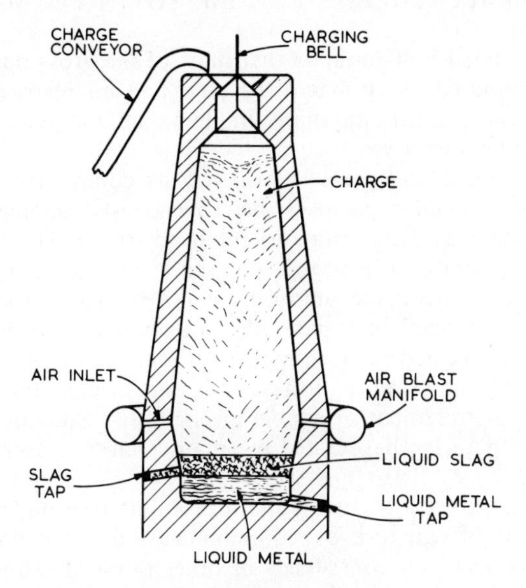

Figure 6.2. Blast furnace

flammable form of carbon (such as coke) and then heated, by burning of the carbon, in an air blast (sometimes enriched with oxygen). As the charge moves slowly down the column the well known reaction takes place

$$Fe_2O_3 + 3CO = 2F_3 + 3CO_2$$

The iron oxide is in the iron ore, the carbon monoxide comes from the coke which as it heats up reacts with oxygen from the air blast and with carbon dioxide released by the breakdown of the limestone. The end result is a rather impure liquid iron at the bottom of the furnace. On the top of the liquid iron floats the liquid slag resulting from the reaction and on top of this is carried the still-solid, mixed, reacting material or *burden* of the furnace. Liquid iron is drawn off in quantity, and taken for further bulk refining into cast irons or steels. Still further refining is likely to be required for more expensive and more special cast irons or steels and the individual quantities treated may fall to charges as small as 50 kg for special purposes.

Bulk treatment makes it possible to produce some finished steels for a cost per kg as little as 1/5000 of that of transistor material at the other end of the scale.

6.8 SIMPLE MEANS FOR IMPROVING A MATERIAL

Having refined a material so that most of the gross impurities have been eliminated many macrostructural and microstructural defects will still remain. Can anything simple be done about this? In many cases the answer is yes.

For example, a material which has its component grains (molecules or embedded particles) heterogeneously ranging in size and orientation may have more useful properties if (1) its grains are made more uniform in size (2) its grains are made uniformly larger (3) its grains are made uniformly smaller (4) its structure is preferentially arranged in a particular direction of orientation (5) the order of its structure is changed, without changing its nature. In many cases one or more of these ends may be achieved by means of simple heat treatment or simple mechanical treatment or a combination of both, perhaps aided by small changes in composition, see Section 6.9.

UNIFORMITY OF STRUCTURE is often desirable in a material, because irregularity of structure is a common cause of mechanical weakness and of variation in consistency of other physical properties. Alternating sizes and orientations of grains, whether crystalline or molecular, lead to uneven behaviour of a material in conditions of externally imposed stress, changed temperature or applied electrical and magnetic fields. This uneven behaviour occurs because the basic structure of almost every material shows some directionality of properties.

Controlled rapid solidification from a molten or vapour state is sometimes an effective means for attaining refinement of structure. On the other hand, heating in the solid state to a suitable elevated temperature followed by controlled cooling can often be used to improve uniformity in a material that shows an easily-reversible allotropic change with rise and fall in temperature (e.g. iron).

Plastic deformation of a suitably-plastic material can be used to fragment the material's structure so that a more uniform structure will form when the material is heated to allow reformation. The process of reformation of structure is particularly relevant to metallic materials in which application it is called *recrystallisation*. When plastic deformation is caused without reformation of structure the process is called *cold working*. Many chain-polymer structures are susceptible to improvement in this way by more orderly alignment induced by plastic deformation in a controlled manner in a particular direction.

GRAIN ENLARGEMENT is sometimes desirable to give greater stıength or more predictable properties. The ultimate is a single large grain if that is possible, as it may be in crystalline materials. Grain enlargement may be accomplished in a suitable material by sustained heating at a suitable temperature in the solid state. This approach is particularly possible with metallic crystals, but single crystals are difficult to grow and are economically impracticable for all except very specialised applications such as solid-state integrated electronic circuits.

Materials usually give better mechanical properties with finer grain sizes but creep properties and some electrical and magnetic properties may improve with increase in the size of the grains.

Since short polymeric chains may tend to pull apart too readily under an applied stress, chain polymeric materials may sometimes be improved in strength by increasing the length of the chains. Chain lengthening is not easy to accomplish after initial synthesis is complete, but it is possible in some materials to induce more cross linking between short chains, perhaps by exposure to heat, or other radiation and thus improve the mechanical properties.

GRAIN OR PARTICLE SIZE REDUCTION is often desirable in materials because the magnitude of most of the mechanical properties seems to be inversely related to the size of embedded particles or grains prevailing within the structure. A typical pattern of variation is indicated in Figure 6.3, which refers particularly to grain size in metals.

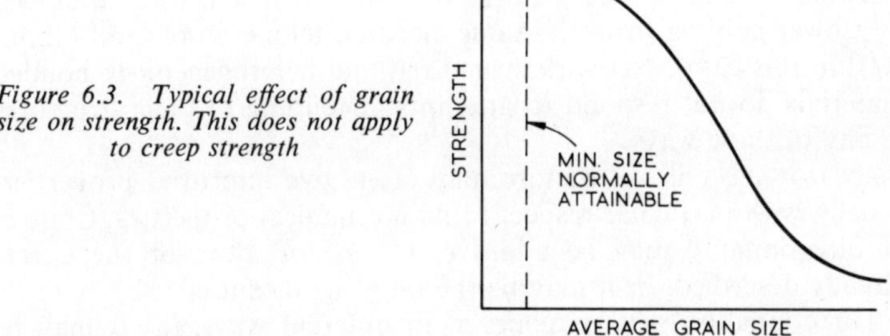

Figure 6.3. Typical effect of grain size on strength. This does not apply to creep strength

It will be noted from the pattern in Figure 6.3 that there tends to be a limiting grain size below which there is little or no gain in strength. Fine grain size may be achieved in most plastic crystalline materials by mechanical deformation of the structure into a strain hardened condition followed by heat treatment at a suitably elevated temperature to recrystallise without causing subsequent grain growth. The mechanism of this operation is that the internal structure is fragmented by the mechanical deformation, each suitable crystal fragment then becoming a potential nucleus for a new grain. On heating the distorted material, reformation of the structure proceeds simultaneously from all the potential nuclei until the whole structure is reformed about the new larger number of smaller-sized grains. Heating beyond this stage gives grain growth as the more stable grains continue to grow by gradually absorbing adjacent less stable ones. It is important that sufficient initial mechanical breakdown of the structure is caused to create a large enough number of new nuclei. If less than about 5% prior deformation is given, then only a limited number of nucleation centres are left and grain growth from such centres can be catastrophically rapid.

Brittle crystalline materials cannot have their grain structure refined by this means, therefore if a fine grain size is required it must be derived during solidification by such means as innoculation. Rapid cooling during solidification is not likely to be a safely effective grain refining process with a brittle material since the chances are that rapid cooling will cause cracking or perhaps *granulation* (fragmentation into individual crystals).

A crystalline material that shows an elevated temperature allotropic transformation may have a coarse grain structure refined by heating it above its transformation temperature and then cooling it down again at a suitable rate. Plain, low-carbon steels can be treated in this way by the process known as *normalising*, which can produce a fine grain structure from a coarse one or reform a work-hardened structure, leaving the material fine-grained and more shock-load-resisting state compared with the fully annealed structure developed by slower cooling from the same elevated temperature level (about 900° in this case). Network structured and heterogeneously bonded materials do not respond to attempts at refining of their structures in any of these ways.

DIRECTIONALITY in a structure may often give improved properties, usually with particular respect to its mechanical properties. Control of directionality may be additive, to one or more of the effects already described, or it may be the only significant effect.

Directionality may be achieved in different ways. (a) It may be developed during production or manufacture of the material by a controlled preferential distribution. (b) It may be achieved by control of the consolidation process, as in controlled directional solidification from the liquid state. (c) It may be developed by an appropriate heat treatment in the solid state in which, by controlled diffusion of its atoms or molecules, the structure is rearranged in a desirable orientation. (d) Materials may sometimes have their structures reorientated by plastic deformation in a controlled direction. The last method is one particularly suitable for many metallic materials and for some chain polymeric materials.

A very simple illustration of the effects of geometric distribution is given in Figure 6.4 in which it can be seen that elongation in one

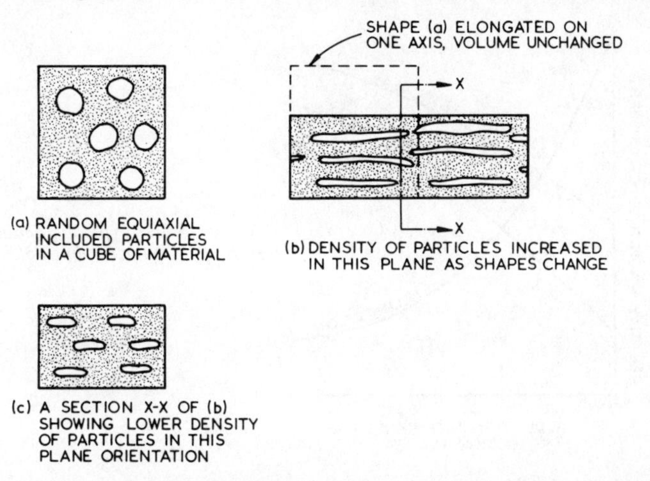

Figure 6.4. (a) Random equiaxial distribution. (b) Density of particles increased in plane of deformation as shape changes. (c) Section XX showing lower particle density

direction leads to an increase in the relative density of embedded particles in the direction of the planes of deformation associated with a decrease of density in the planes transverse to the direction of transformation. A macrostructure having a detectable directional distribution is said to have a *fibre* structure.

CHANGE OF ORDER IN A MICROSTRUCTURE, particularly a crystalline structure, is sometimes a useful means for influencing properties, particularly mechanical properties. The commonest method is to use cold working to cause *work hardening*, particularly in plastic metallic materials. In this context the purpose of applying the deformation is not to induce directionality but to cause some disorder in the crystalline structure by fragmenting the individual grains so that the continuity of the crystallographic planes is disturbed. This disturbance breaks some of the interatomic bonds, usually an undesirable effect, but causes interlocking of dislocations within the structure inhibiting some of their effects, raising the yield strength and giving an apparent improvement in the mechanical properties associated with some decrease in plasticity. The decrease in plasticity occurs because some plasticity is used up in causing the work hardening. Material in the work-hardened state is sensitive to elevated temperature, which can initiate softening and recrystallisation and so start a reversion of properties back to their original values. A typical pattern of variation of mechanical properties with degree of deformation is shown in Figure 6.5.

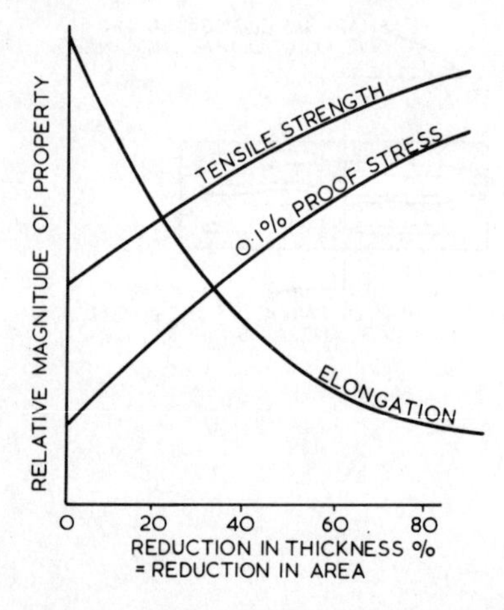

Figure 6.5. Typical changes in properties induced by cold rolling

Other changes of order are possible in different materials and can have different effects. It is sometimes possible to change the interlinking order in a chain polymeric material and so change the mechanical properties of the material without affecting its basic nature. Some materials are polymorphic and the nature of the interlinking of the constituent atoms or molecules may be changed with marked effects on their properties. Carbon has already been mentioned in this connection, but this kind of change if carried far enough leads to a complete change in structure which takes it outside the scope of the present consideration.

6.9 USE OF MINOR CHANGES IN COMPOSITION

Some of the means for controlling properties may be more readily applied if assistance is given by chemical means. This particular consideration does not concern those chemical means which radically alter the structure of a material but only those which aid the application of the simple methods. Three main effects can be aimed at in making minor chemical changes

(1) change in the geometry of microstructural distribution
(2) change in the nature of impurities unavoidably present
(3) inhibition of the effects of certain microstructural defects

The first depends predominantly on nucleation effects caused by minute additions of suitable innoculants. Thus, the production of a fine-sized, uniform grain structure within a phase solidifying from a liquid state may be facilitated by the presence of a small quantity of a foreign element or compound which will act as a solidification nucleant. As little as one part in ten thousand can be sufficient to produce an optimum refinement of structure in some cases, but the amount depends on the materials involved and also on the nature of the solidification process. As a rule, the refinement of structure induced by innoculation causes also a finer dispersion of impurities, dissolved constituents and precipitated phases.

Many examples of this kind of treatment are found in metallic materials of construction in which the first stage of fabrication is usually a casting process involving freezing of a liquid metal or alloy as it cools down from an elevated-temperature molten state. Probably, the best known example is the innoculation of grey cast iron (a heterogeneous brittle material with low tensile strength) with small quantities of such materials as cerium or magnesium, to produce spheroidal graphite (s.g.) iron which has a much finer, more uniform structure with significantly higher tensile strength and some degree of ductility (from about 285 N/mm^2 upwards and about 15% El).

The second aspect of minor chemical change, namely that of change in the nature of impurities, can be a very important one in many situations and particularly when an unavoidable impurity has a marked effect on properties. Probably, the worst situation arises when an impurity, perhaps present only in small amounts, distributes itself as a continuous network film within the body of the parent material, completely partitioning grains, phases, or molecules, or making the bonding between them much weaker and thus affecting almost every property.

If the nature of the impurity can be modified by reaction with an added constituent one or both of two possibilities may result (a) the mode of distribution may be changed to a less harmful form (b) bonding characteristics of the modified impurity may be changed to stronger less harmful types than those of the unmodified impurity. It is sometimes possible to change a continuous film-type distribution into a dispersed type with less-harmful isolated islands of impurity or to change an angular type of impurity that causes stress concentration to a rounded type with less harmful effects. Simple examples can be found in metallic materials. Sulphur is a normal impurity in plain low carbon steel and tends to form a continuous brittle grain boundary film of iron sulphide which can have seriously harmful effects on the mechanical properties of the steel. If while the steel is molten some manganese (1 %) is added it forms manganese sulphide by preferential combination with the sulphur and this compound forms into relatively harmless globular islands when solidification takes place. Gases dissolved in a material, possibly by reaction in the course of a manufacturing process, may sometimes have their harmful effects limited by reaction with a suitable added agent to form a relatively harmless dispersed compound.

Use may often be made of the third possible effect of a minor chemical change to inhibit or minimise the effects of some inherent microstructural defect. For example, microfissures may sometimes be avoided by adding a constituent which prevents the development of fissures by modifying their formation process. Degassing or gas-fixing agents often have this effect as the primary purpose of their action. Another possible purpose of a suitable additive may be to lock dislocations within a structure (commonly applied to plastic crystalline types of structure) so that the yield stress is raised. This end is attained by using an element that will dissolve in the matrix and diffuse preferentially into the zones of distortion associated with dislocation defects stabilising the dislocations and making them harder to move.

6.10 USE OF MAJOR CHANGES IN COMPOSITION

The means for control of properties which have been outlined so far are all simple means which do not greatly influence the basic structure of a particular material. Now we must consider compositional changes that do cause radical or major changes in the behaviour. Two things should be noted.

(1) There is no one fixed limit of composition change below which the effect of an additive is minor and above which the effect is major. Each material is different from every other and each prospective addition has its own characteristics.

(2) The principles governing simple treatments can usually be applied also to materials which have been subject to major changes and may have effects additive to the major effects.

Major changes in the structural behaviour of a material when another material is added to it are dependent on the following factors

(a) degree and type of solubility of the addition in the parent material or vice versa

(b) the nature of any change in mutual solubility, particularly solid-solubility, with change in temperature
(c) the degree of metastability of any super-saturated solid-solution that may form between them
(d) the nature and behaviour of any compound, or compounds, that may form between the materials

Such factors may operate in differing combinations with different materials, the extent and types of each combination determining the nature of the major change it induces. Before considering practical applications it is necessary to summarise what is relevant to each factor.

DEGREE AND TYPE OF SOLUBILITY of one material in another is a fundamental consideration in almost every application of material science. Between insolubility and total solubility are many possible alternatives some of which are of little practical use and others which are the foundation of many useful materials. Nearly all metallic materials and a very wide range of non metallic materials derive their properties from one or other form of solid-solubility relationship.

There are two basic forms of solubility: *Substitutional solubility* in which atoms or molecules of the solute move into places in the structure of the solvent material normally occupied by atoms or molecules of a different kind, displacing or ejecting the latter. Probably most commonly found in crystalline materials. *Interstitial solubility* in which atoms or molecules of the solute move into suitable gaps or interstices in the structure of the solvent. Materials with network structures are particularly likely to form solvents suited to this kind of relationship. Very small atoms are always potential solutes of this kind which explains why elements such as hydrogen, boron and carbon are often found as interstitial solutes.

In all cases of solubility there must be some affinity between the atoms or molecules of the solvent and the atoms or molecules of the solute if the relationship is to exist at all. Thus completely stable atoms such as helium and argon are not found in solutions of any kind except under unusually extreme environmental conditions.

A state of solubility may have a high degree of structural order and stability but is essentially a transitory state that can be changed by temperature. For this reason a particular solution of two or more substances differs from a compound which has a fixed composition and behaves as an independent permanent substance.

Solid solution affects the structure of the solvent. The overall result must be some increase in inter-atomic stability or the solution would not take place, but the effects on individual solvents differ widely with the type of bond relationship between solvent and solute. For example, one of the most obvious effects is that a solvent's normal solid structure will be distorted by the presence of a solute. If this strain energy becomes excessive, with build-up in quantity of solute, a proportion of solute will be reached at which the increase in strain energy offsets the increase in stability and solution stops. This means there is likely to be a specific limit to the solubility of one substance in another.

Distortion of structure can be particularly important to the mechanical properties of a material. Any type of structure is liable to be mechanically weakened by solution distortion unless the result of the solution is either to cause (a) a reordering of the structure into a more stable array (b) to create interconnecting bonds where none previously existed (c) to bolster up an inherently weak array. Each effect is possible with differing materials. Reordering of structure is possible in many crystalline arrays and some network structures. Additional interconnecting bonds may often be created by solution in polymeric materials, and gels may sometimes be stiffened by absorption of a suitable interstitial solute.

Apart from solid solution influencing properties of materials, it can also form the basis of the joining of one suitable substance to another, as in the *sintering* together of the dissimilar particles in a compacted aggregate to form a heterogeneous solid, or in the use of fusion welding, or solid phase welding, to bond a bulk of one material to a bulk of another (processes of this kind are outlined in Volume 2) by partial diffusion of one substance with the other.

CHANGE IN SOLUBILITY is probably the most potent tool in the hands of a materials technologist, particularly when the change in relation to particular materials is temperature-dependent and in the solid state. It is certainly the principal means used for improving the mechanical properties of those metallic materials which can be supersaturated and then are metastable at the required working temperature.

The most obvious change in solubility is likely to be that which occurs during the freezing of a thermally melted substance. Solutes usually dissolve more readily in a liquid (diffusion is more rapid) and in the majority of cases in which full mutual solubility is not possible a greater proportion of a particular solute is likely to dissolve in the liquid state of a solvent than in the solid state of the same solvent.

An abrupt change of this kind occurs in a *eutectic reaction* where at a particular level of falling temperature, a specific composition of molten solution solidifies and simultaneously separates into its separate solid constituents, the substances appearing side by side in the structure of the solid (see Plate 1) in a distribution that depends on the nature of the constituents and on the mode of cooling. The reverse happens in reheating. Thus two things may become possible with materials of this kind that can be melted readily: (1) easy mixing can be achieved by melting such liquid-soluble constituents together (2) some degree of fine solid mixing of otherwise mutually insoluble solid materials may be accomplished by making use of any limited liquid solubility they may possess to blend the constituents, and then trapping them in a solid mixed state by freezing so rapidly that at least some of the solute is trapped in a network of the solvent matrix.

The reverse type of change to a eutectic reaction called a *peritectic reaction* can occur in many materials—this cannot be used to induce supersaturation but may be a means for producing a stable solid solution structure.

Greater control of properties is possible in a material that shows a change of solid solubility with change in temperature, notably a decrease of solubility with fall in temperature. When change of this kind takes place in the solid state of a material diffusion rates are likely to be fairly slow and this may make it possible to exercise precise control over the composition and state of the structure at any given stage, particularly if the degree of metastability of the structure is high.

The most abrupt change in solid solubility is likely to occur during an allotropic change in which a state of complete solubility of a given proportion of a solute in one state of the solvent structure may be abruptly changed to substantial or complete rejection in the other state. If the lower temperature state is the one with lowest solubility, the effect can be the solid equivalent of a eutectic reaction and is called a *eutectoid* reaction (see Plate 2a). In the latter reaction, at a particular level of falling temperature, a specific proportionate composition of a solid solution precipitates out into its separate constituents which are left closely intermingled in each other and, usually, firmly bonded to each other in an integrated structure.

There is also a form of reaction in the solid state, called the *peritectoid* reaction which is similar to the peritectic reaction. It is the reverse of the eutectoid reaction in that two separate solid phases present in specific proportions in a solid material go into solid solution in each other at a particular level of falling temperature.

Many of these solid-state reactions when they occur in plastic crystalline materials such as metals or alloys can cause severe distortion of the parent lattice, greatly influencing the latter's mechanical properties by the inhibiting effect of the distortion on the movement of dislocations. Probably the most outstanding distortion can be caused by a eutectoid type of reaction in which a former solute is most likely to be trapped in a state of metastable supersaturation and be artificially held in the structure of the former solvent. The severity of this kind of distortion depends on the metastability of the structure in relation to the degree of supersaturation.

DEGREE OF METASTABILITY is the determining factor with respect to the durability of any condition of instability. Thus any treatment that makes use of a solubility change to develop desired unstable structural conditions in a material (such as a state of supersaturation) can only be effective if the conditions of subsequent treatment or subsequent service are such that the structure does not revert too quickly back towards its stable state. Hence precipitation treatments (see below) much used in metallic alloys, depend both on the degree of supersaturation that can be developed by the treatment and on the resistance of the metallic crystalline lattice to stabilising diffusion of atoms or molecules out of their unstable positions.

FORMATION OF COMPOUNDS can be a potent factor in the control of the structures of many materials. Sometimes the presence of a compound within a material is helpful by increasing internal bonding and/or geometric interlocking but, on occasions it is a nuisance because it uses up bonds which could be more usefully employed. (Burning of flammable material by absorption of oxygen is an example of this when disintegration occurs as the oxygen takes up bonds previously used within the structure.) Between these modes of behaviour are many alternatives.

A compound is a relatively permanent combination of two or more basic substances (see Table 5) which cannot be separated by simple means. Therefore the molecules of a compound can behave like the atoms or molecules of elemental substances, forming solubility relationships and/or reacting in different ways. If a compound such as a metal oxide can be finely dispersed through the structure of a metallic material, a hardening called *dispersion hardening* can occur which is similar in effect to precipitation hardening mentioned later in this section.

If a compound is wanted in the structure of a complex material it is likely that it will be prepared separately and added at a convenient time in the proportions required. Alternatively the separate constituents of a compound may be added at some stage in the preparation of the main material, perhaps going into solution with other constituents and then caused to react within the material to produce the required compound. A reaction of the latter kind may occur spontaneously when the correct proportions of constituents are present in association with each other at some particular stage of the treatment, perhaps by heat alone, perhaps during a change of solubility that takes place during heating or cooling.

Spontaneous reactions of this kind may happen unintentionally if a material containing the potential constituents of a compound is maltreated in some way, perhaps by overheating or subjecting it to prolonged undesirable irradiation. The effect may be to make a complete change, usually harmful, in the properties of the original material. Ultraviolet and other forms of radiation, notably atomic radiation or even light, in some cases can cause marked changes in this way in a wide range of materials. This kind of change should not be confused with other changes which irradiation may also cause, such as disturbance of the pattern of distribution of, or the stability of the normal bonding between the constituents in the material.

Each or all of these phenomena may be used with different basic materials to develop new or modified materials with properties suited to specific purposes. Some phenomena are more important than others with respect to particular types of materials. Thus control of solubility is of major importance to the majority of metallic materials and compound formation and control are the main factors in network-structured materials and many polymeric materials. Strengthening of the structure of a metallic crystalline material is likely to depend on inhibiting and controlling the effects of dislocations since these effects are the most usual sources of the greatest weaknesses in metallic material. Control depends to a large extent on causing distortion and irregularities within the structure which will both limit the numbers of dislocations and hamper their movements. In other types of materials much more will depend on achieving increased density of interatomic, or intermolecular, bonding and/or more systematic arrangements of bonds, perhaps by forming suitably distributed intermediate bridging compounds.

6.11 USE OF SOLUBILITY CHANGES IN ALLOYING

Solubility relationships are very important to many kinds of materials but the most versatile uses are found in metallic materials.

Very few metals are used in their pure elemental form except for special purposes. They are usually mechanically weak and also soft and plastic. 'Commercial purity' metals are not significantly stronger than pure metals and the only mechanisms suitable for strengthening them are grain refinement and work hardening, neither of which are suitable for many purposes. It is only when properties other than strength are of particular importance that elemental metals are used, as for example in the use of copper or aluminium for electrical power conductors where low resistivity is desirable.

Metals are particularly suitable for the kinds of treatment that make use of solid solution conditions and controlled change in solid solubility. This suitability arises from three causes. (1) Metals have the potentiality of high ideal strengths inherent in the orderliness of their structure and in the high density of the moderately-strong bonds that can form in them. (2) The continuity and elasticity of the normal orderly metal structure which often makes solid-solution combinations of suitable constituents possible and sometimes permits effective control of potential changes in solubility. (3) The amenability of dislocations to control by obstructions of various kinds.

Apart from the methods outlined in Section 6.8 improvement may be achieved by *alloying*, that is by adding one or more constituents to the metal structure. An *alloy* may appear to be a radically different material but it must still retain the basic characteristics of the parent metal. Non metallic materials may be used as alloy additions, provided that the last condition is not violated. Alloying may be no more than the addition of a trace element to aid solidification-nucleation and refine the grain size, but in most cases it is more than this. Thus an alloy addition may (1) control an undesirable but unavoidable impurity (2) improve existing properties, notably strength (3) impart new physical or chemical properties.

The final effects of alloying a particular metal will depend on (a) the nature of the alloying additions (b) the proportion of alloys added (c) the basic effects of alloying on the structure (d) the subsequent treatment given to the structure after alloying.

Stable solid solutions can be very useful for alloys, the most stable usually being substitutional solid solutions. Substitutional solid solutions are able to form only when the differing types of atoms concerned (or molecules in the case of molecular materials) arc suited to each other, both by being able to form similar bonds, or bonds with nearly similar geometric effects, and by being of a near enough equivalent atomic size to fit into each other's lattice without causing too much distortion. If a solution relationship is likely to cause too great a distortion in a structure the resultant strain energy in the system will be sufficient to cause surplus would-be solute atoms to diffuse away to an outer surface or to the boundary of the phase. If the atomic fit is sufficiently good, complete mutual solid solubility may be possible in all proportions at all temperatures. As fit and other properties become less compatible the possibility of substitutional solid solution between differing metals, or metals and other elements, becomes less and less. As match decreases the proportions of one material likely to dissolve in another become smaller and likely to exist over smaller temperature ranges. On the other hand if a differing bond type begins to predominate in the presence of an excess of a particular solute, a solvent's structure may begin to change its basic order to accommodate the additional solute. For example at n.t.p. up to 39% of zinc will dissolve in copper by forming a substitutional solid solution in the FCC lattice of the copper, imparting considerable improvement in elastic strength and improving the forming properties but also increasing resistivity. Such an alloy of zinc in copper is known as a *brass* and several alloy compositions are in common use, see Figure 6.6. If 39% Zn is

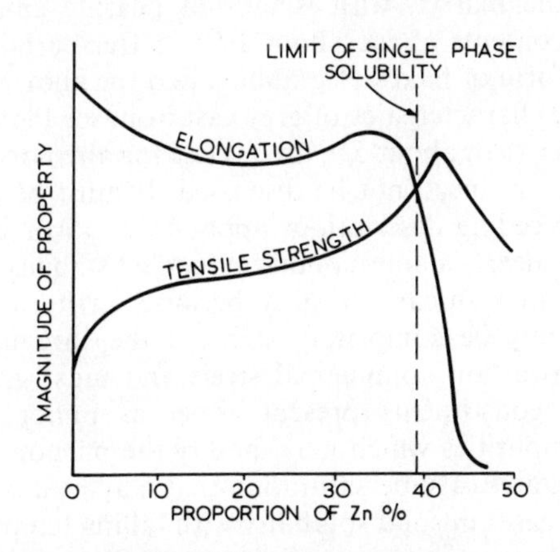

Figure 6.6. Change in mechanical properties of single phase copper-zinc solid solution with change of zinc content

exceeded at n.t.p., the single-phase solid-solution structure begins to contain a very brittle second phase, called β brass, which has a BCC lattice containing equal numbers of copper and zinc atoms. At higher proportions of zinc, particularly above 46%, this and other changes that occur make the alloys useless for most practical purposes.

Stable interstitial solid solutions are much less likely to form than are substitutional solid solutions. Because the solute atom for this kind of relationship must be very much smaller in size than the parent solvent atoms, the range of compositions over which solution is likely to exist must be limited and mutual solubility impossible. When such solutions do exist the parent lattice is likely to be severely distorted. Severe distortion can give marked improvement to some of the mechanical properties of a structure but is also likely to cause such a sharp reduction in plasticity that embrittlement is probable. Interstitial solubility is likely to decrease with decrease in temperature and vice versa. To see the limitations one need only consider the iron-carbon system as found in steels. At n.t.p. only about 0·01% C will dissolve as an interstitial constituent in iron (forming the phase called *ferrite* commonly found in steels). If the carbon content is increased, a new phase compound of iron and carbon (FeC) called *cementite*, insoluble in the ferrite at n.t.p., is found embedded in the ferrite. Ferrite and cementite usually appear side by side in localised areas closely intermingled in fine alternating lamella in a state called *pearlite*, see Plate 2a. At about 0·8 % C the whole structure is in this form. With larger amounts of carbon, cementite begins to predominate as the matrix, with islands of pearlite embedded in it. With carbon contents above about 1·5 % a free carbon phase may appear in the form of flakes of graphite, and the alloy begins to take on some of the characteristics of grey cast iron, see Plate 2b. Carbon contents greater than about 5 % are useless for any normal constructional purpose and need not be discussed. It must be kept in mind that (a) the preceding descriptions apply only to iron-carbon alloys in an annealed, nearly stable, condition (perfect stability is impossible to achieve in an iron-carbon alloy because ferrite, cementite and graphite can only be completely stable if they are unmixed) and (b) most plain-carbon commercial steels and most grey cast irons contain other constituents present either as minor additions or unavoidable impurities which may modify the phenomena.

Metastable states of supersaturation that may be derived from the control of decreases in solid solubility with falling temperature, form some of the most promising conditions for giving high strength to an alloy. Such states are potentially able to cause the maximum tolerable amount of distortion in a given lattice and hence the maximum hindrance to dislocation movement. This hindrance in conjunction with inhibition of dislocation sources, can raise the level of elastic properties and stress for final failure to a surprisingly high value for a number of alloys.

The degree of distortion obtainable in a particular situation depends on the rigidity of the parent lattice relative to the number and

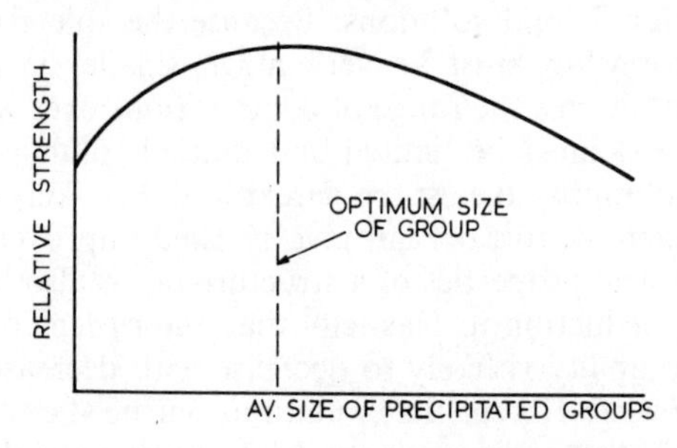

Figure 6.7. Typical effect on strength of size of precipitated groups

mode of distribution of the trapped atoms. With the majority of suitable alloys it is not fine dispersion of isolated atoms that gives maximum strengthening but the precipitation and coalescence of the entrapped atoms (Figure 6.7).

Thus we get the concept of *precipitation hardening* which is a three-part treatment (1) a *solution treatment* at elevated temperature to dissolve the solute (2) a *quenching* operation to trap the solute as solubility decreases (3) a *precipitation* or *ageing* treatment to develop the required size of precipitate. For such a treatment to be possible and effective there are several requirements.

(1) The solid solubility of the solute in the given alloy must increase markedly with rise in temperature above n.t.p. and decrease equally definitely with falling temperature as shown in Figure 6.8a.

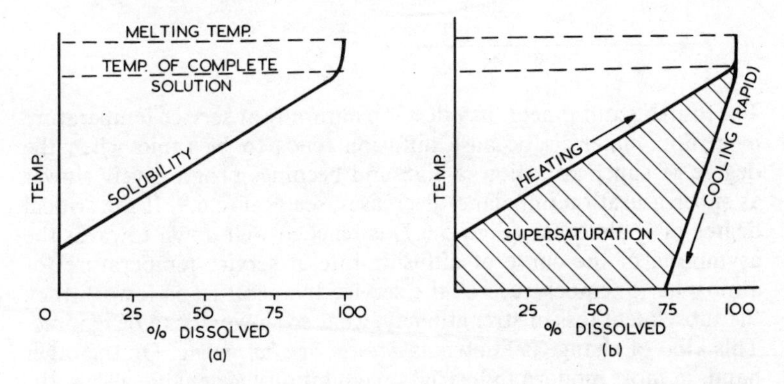

Figure 6.8. (a) Change of solution with temperature of a given amount of potential solute in a solvent. (b) Rapid cooling from full solution entrapping some solute

(2) The rate of diffusion of rejected solute out of the solvent matrix, when the temperature falls below that suitable for a state of complete solution, must be so slow that the rejected solute can be trapped in a supersaturated state in the matrix by rapid cooling at a reasonably-attainable rate, see Figure 6.8b.

(3) Once cooled down to a room temperature or a temperature-level appropriate to the material (some complex alloys may require more than one stage in the treatment if more than one constituent is involved in the changing solubility), the structure must be sufficiently metastable to prevent further undesirable changes in the state of supersaturation within a reasonable time after quenching.

(4) The rate of diffusion of entrapped solute atoms (or molecules) to coalesce with each other must be controllable so that the critical amount of coalescence of precipitated units may be developed, but subsequent change must be so slow that the material has a useful service life in the desired state.

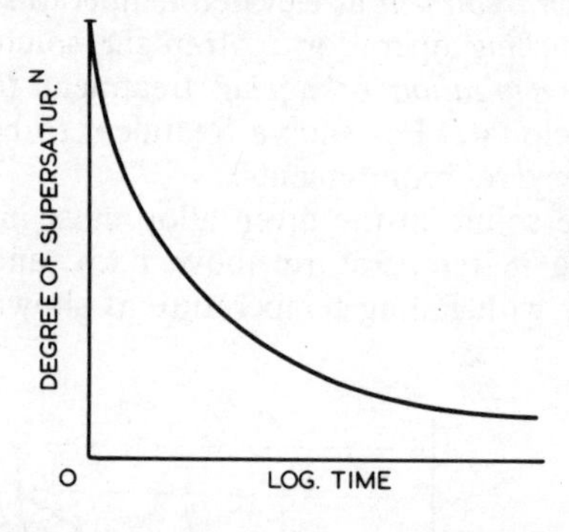

Figure 6.9. Typical ageing curve for a supersaturated solid solution. The ambient temperature influences the curve shape

The fourth requirement may develop naturally at service temperature in certain materials because diffusion tends to be rapid when the degree of supersaturation is high and becomes progressively slower as supersaturation imbalance decreases, see Figure 6.9. If the critical degree of coalescence (Figure 6.7) is reached well down towards the asymptote of the curve of diffusion-rate at service temperature (or atmospheric temperature of the service temperature is lower), then the subsequent loss of strength may occur so slowly it can be ignored. This kind of change is known as *natural age hardening*. On the other hand, in most modern industrial precipitation-hardenable alloys, the material in the solution-treated state is so metastable at or below service temperature, that there is no measurable change in coalescence within a long period of time. Therefore the critical size of coalescent groups must be developed by treatment at a temperature high enough to give the required degree of controlled coalescence within a reasonably finite time. This time must be long enough for control to be reasonably easy and accurate, but not so long that the process becomes uneconomic. Such treatment is known as *precipitation treatment* or *artificial ageing*.

The classic example of these treatments was found in Duralumin alloy (4% Cu in aluminium) which gave a natural-ageing supersaturated solid solution, after quenching from a solution treatment temperature of about 500°C, and led to the development of a whole range of naturally and artificially aged aluminium alloys. Many alloys of other metals have been developed for their response to precipitation hardening of this kind, including some of the more complex steels, although the principal mechanism of hardening of steel is, more commonly, of the special kind described below.

The hardening of steel is usually based on a modified precipitation behaviour called *quench hardening* in which the maximum distortion of the parent lattice is achieved immediately after quenching from a suitable solution temperature. This maximum hardening occurs by virtue of the entrapment of atomically dispersed interstitial carbon atoms in a lattice in which they fit very badly indeed. Effective treatment of this kind is applicable, most commonly, to steels in the medium to higher carbon class. The effect is caused by the change from FCC to BCC which occurs in a steel as it cools down from elevated temperature. The temperature at which the change actually occurs is usually below 800°C and depends both on the alloy composition and the rate of cooling of the steel. The extreme hardness arises because all of the carbon in these steels is soluble in the FCC structure (*austenite*) but is almost completely insoluble in the BCC structure (ferrite). When a steel is heated up to its transformation temperature its cementite dissolves in austenite and the carbon atoms disperse uniformly through the matrix lattice interstices. In cooling, three things try to happen simultaneously

(a) austenite (or γ iron as it may also be called) tries to change back to ferrite (α iron)
(b) carbon atoms try to get out of the transforming lattice
(c) carbon atoms try to combine with iron atoms to form cementite.

If these changes are forced to try to take place too rapidly they hamper each other and can be delayed to a much lowered temperature level where the lattice becomes more rigid (because of the loss of thermal softening effects) and a severely-distorted brittle-hard metastable state, called *martensite*, is retained.

Martensite is a body-centred tetragonal structure (see Figure 3.14) in which the carbon atoms are held in supersaturated dispersion and, although it is very hard (its hardness may be over 700 HV) and seemingly strong (tensile strength may be over 2 kN/mm^2) it is brittle and liable to shatter under impact. In this form martensite is useless except for a very few purposes, so it has to be *tempered*, that is reheated to a suitably raised temperature at which some of the suppressed changes may develop a stage further, reducing the hardness and strength but increasing the ductility to a desirable level, as a *tempered martensite* structure is developed (see Plate 3). The effect of differing tempering temperatures on one quench-hardened steel is indicated in Figure 6.10. Reheating above a minimum tempering temperature, appropriate to the particular steel, causes reappearance of pearlite (Plate 2a) in a very fine form called *bainite*.

Plain carbon steels require a very rapid cooling rate to obtain good hardening. Therefore because the rate of cooling must get slower towards the centre of any section that is cooling simultaneously from both sides, there is a fairly low upper limit to the maximum thickness that can be hardened right through to its centre. There is

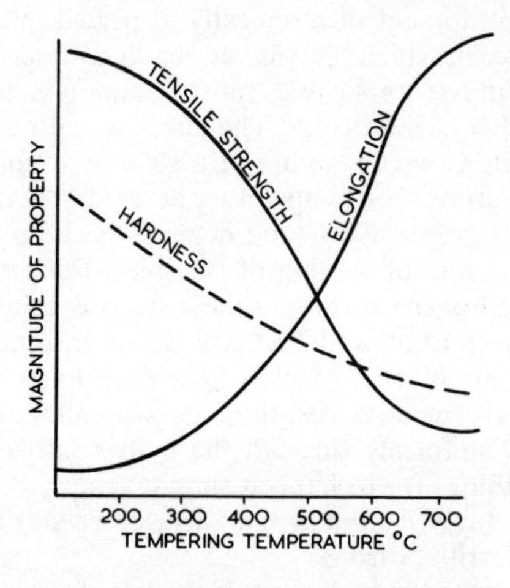

Figure 6.10. Patterns of change in mechanical properties of a quench hardened steel with tempering temperature

usually a limiting rate of cooling from the outside for any given steel, beyond which it is impractical to go because too-rapid contraction from the outside may cause *quench-cracking*. It is possible to modify this limiting behaviour of quench-hardenable steel by making alloy additions such as nickel and chromium which slow down and modify the atomic changes and lower the critical rate of cooling. Lowering the critical rate permits the hardening of thicker sections and can reduce the thermal stresses set up by a critical cooling operation.

The structural change from FCC to BCC involves an expansion of the material lattice which opposes normal contraction behaviour during cooling. Taken in conjunction with uneven thermal contraction this can cause distortion in the shape of a steel sample and can add to the risk of causing quench cracking. Thus effective quench-hardening treatment may imply not only that there are limitations on the shape and size of the mass being treated but also that there may be a necessity for making some small allowance of extra material to be machined off after treatment to bring the final shape to the accuracy required.

6.12 SIMULTANEOUS USE OF SEPARATE MATERIALS

One method which is still in common use for gaining the combined advantages of different materials is simultaneous use of two or more materials in separate forms to gain the best services from each.

For example, wood is often used as a structural material but it has low resistance to atmospheric conditions, so the wood is painted. The wood gives the strength while the paint gives the resistance.

Another application is reinforced concrete in which concrete, with a high compression strength but low tensile strength, is reinforced

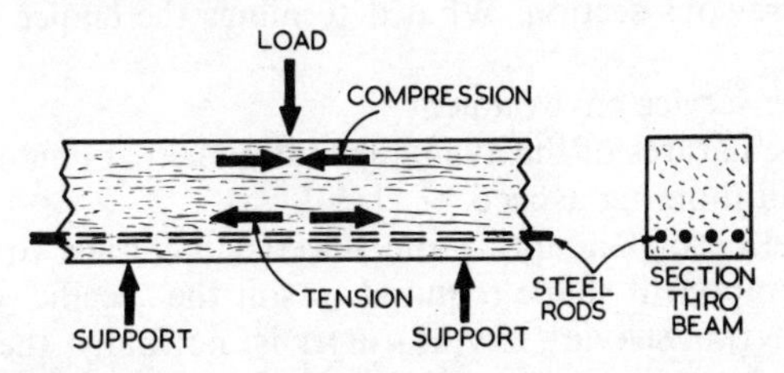

Figure 6.11. Reinforced concrete with embedded steel rods to absorb tensile loading

with steel bars of relatively high tensile strength, suitably embedded in position to take the main tensile load, as suggested in Figure 6.11 for a reinforced concrete beam.

Metal may also be used with metal. Two examples can be used to illustrate this. Overhead electric power transmission lines require low resistivity and high strength, so they may be made with a high strength steel core which takes the load, and a close-fitting, thick-walled, aluminium shell, which carries nearly all the power and is light in weight but adds little to the strength. The other example is the use of a thin layer of a strong, corrosion and temperature-resisting metal for lining the inside of a chemical-treatment vessel, the main part of which is made of a cheap, strong, high-strength low-alloy steel (a low-alloy steel is one with less than about 3% by weight total alloy addition).

It will be noted that there are three possible modes of simultaneous use (1) the different materials are entirely separate from each other in separate parts which fit together to make up a composite assembly, (2) the materials are substantially separate units bonded to each other along continuous common faces (3) one material is embedded as a prefabricated unit into a matrix of the other.

The essential feature is that each material has a distinctive integral bulk with an overall geometric shape determined primarily by the geometry of the application and not by any bonding relation it may have with its associate material. That is, geometry of the material is tailored to the particular application. In this lies the difference between the present class of use and the composite materials considered in Section 6.14.

6.13 FACTORS CONTROLLING THE SIMULTANEOUS USE OF SEPARATE MATERIALS

Three modes of simultaneous use of separate materials were listed in the previous section. What determines the choice of a particular mode?

(1) the service environment
(2) the natures of the available materials that could be adpated to suit differing aspects of (1)
(3) the cost of adapting the selected materials to the particular component shape required to suit the specific service.

THE ESSENTIAL SERVICE ENVIRONMENT is, normally, the first factor to be considered in any application of a material and this factor falls into a number of subdivisions.

(1) chemistry of the local environment
(2) environmental pressure
(3) service loading conditions
(4) environmental temperature.

Note that these considerations are not quoted in an invariable order of priority. The specific order of importance will depend on each particular situation, but in *every* case each aspect should be carefully considered at some stage. Note that possible variation in conditions should *always* be taken into account both with respect to frequency and duration. From these conditions the minimum essential combination of properties required from suitable service materials must be determined.

THE NATURE OF THE AVAILABLE MATERIALS determines how and when each can be used. Of the available range each could be suitable in some respects but only a very small number will be completely suitable in all chemical and physical properties and the chances are that these very materials will be too costly to use in bulk or are unsuited for other reasons.

The next step then is to see if the relevant local construction can be adapted to make simultaneous use of separated materials.

THE COST OF SHAPING A MATERIAL to suite a particular service is a major factor in most practical applications. Many subsidiary influences can affect this cost. For example, the material may be so inherently hard and/or brittle that it can neither be machined nor plastically formed in any way. Also, even if a brittle material is capable of being melted, it is probable that it will not solidify without cracking after being poured into a mould. Even if a material is not brittle, its melting temperature may be so high that it cannot be melted, or it may be so viscous in the molten state that it will not flow into a required shape (the complexity of shape that is required greatly influences the level of fluidity for any operation of this kind).

Taking these influences into account it may be found that such complexity of fabrication and so much manufacturing time is involved, in a particular case, that costs are too high. Hence the only way in which the particular material can be used as an independent material is as a thin film, perhaps chemically deposited or electrodeposited, on a suitable supporting surface. Chromium is an example of a material of this kind used mainly as a protective electrodeposited coating. It is a metal that it is almost impossible to use independently in any other way because of the difficulties in controlling embrittling impurities. On the other hand, some materials such as enamel paints are likely to be so physically weak and difficult to produce in solid bulk form that they cannot be used other than in relatively thin films brushed or sprayed into place.

It is sometimes possible to produce large quantities of a particular material, difficult to produce in small quantities, in basic simple forms at low cost, because the material is processed on continuous production lines. Thus with relatively simple subsequent manipulation of basic form, or by using two or more basic forms in conjunction, a useful assembly might be made up. Common basic forms are plates (relatively rigid sheets cut to standard size), sheet (flexible, thin material, usually of standard widths of long lengths wound on to a mandrel or bolster, e.g. paper and cloth), wire (thin and flexible enough to coil) rod (restricted lengths too stiff and/or large in cross-section to coil), section (long prismatic lengths of usefully cross-sectioned shape), and hollow sections such as tubes. Basic forms of a material can be particularly useful if the material can be bonded into unit constructions by glueing or welding.

Materials such as wood, with its marked directionality of structure, its ease of cutting, and its adaptability to bonding with synthetic-resin glue, is particularly suitable for assembly treatment of this kind and has led to the development and common use of the laminated sheets known as plywood. In plywood each thin individual sheet (or ply) has its fibre structure lying parallel to the plane of the sheet and successive sheets have their fibres laid in differing orientations, usually alternately at right angles, before the sheets are glued together. It is not necessary for all the sheets to be of the same kind of wood and inner sheets may be made of cheap softwood and only the outer sheets of more expensive hardwood needed to give good wear and appearance. Quite massive total thicknesses may be built up but, because of limitation of demand, about 25 mm is the normal maximum. Because wood in single ply form is pliable it can often be flexed into quite sharply curving shapes on suitable formers and then held there as successive plies are glued on to form a strong laminated structure with a required contour and directionality of structure. Similar techniques are applicable to other natural and many synthetic materials when there is sufficient demand for the product. Interleaving sheets for such uses can be membraneous in form or they may be woven. Indeed a woven sheet is one way of adapting the strongly directional properties of a thread, yarn or cord (see Figure 2.13) of small approximately round cross-section, but great length, to the needs of a sheet which it might not be possible to produce by other means.

Manufacture of laminations in these ways is not limited to one type of material and various composites made up of mixed laminations of wood, synthetic materials and metals are made for particular purposes. One simple example is the glueing of two strips of aluminium foil alternating with strips of impregnated paper. This composite can then be rolled into suitable sizes to form the basis of a paper capacitor for electric purposes. Other materials may be used in similar ways for other purposes.

An example of a completely different form of composite is insulated conductor wires for electrical installations. Long lengths of copper wire are made and then continuously coated with a suitable thin shell of insulating material, such as polyvinyl chloride (PVC). In this case the shell is not bonded to the wire (it must be readily removable for preparing end connections). The main function of the coating is electrical insulation, but it also protects the wire from corrosion and from some risk of abrasive damage.

Metals are often prepared with other metals. There are many examples of these including tinplate (steel plate coated with a tin alloy), galvanised sheet (steel plate coated with zinc). In each of these examples the materials are firmly and uniformly bonded to each other by controlled local interdiffusion, developed by depositing the coating as a liquid, or by electrolysis on the surfaces of the plates, or by solid-state diffusion bonding (solid-phase welding) of one plate to the other, whichever is cheapest. In each of these basic composites the strength of the inner plate is intended to be used under the blanket protection of the particular coating against specific forms of corrosion.

Embedding of one material in another tends to be a more specialised simultaneous use of materials and rather beyond the present context, but there are one or two examples of basic composite materials in this class. These are fabric or wire reinforced synthetic sheets, wired glass etc. Note that the difference between this class of material and the impregnated fabric materials already mentioned is that in the latter the fibrous material predominates in bulk, that is the fibres are little more than coated whereas in this case the bulk of the material is in the matrix and the cords or wires are embedded within it and more or less completely separated from each other.

6.14 INTEGRATED COMPOSITE MATERIALS

This chapter has already considered the mixing of differing elements and compounds into one coherent material and the simultaneous use of different materials in specific bulk form. Between these two extremes there is a wide range of materials called *composite materials.*

The materials in this range are *integrated composite materials* where two or more basic materials with different characteristics are closely intermingled but distinguishable. At one end of this particular range of composites are the reinforced materials in which at least one of the basic materials is prepared initially in a more or less finished massive form and then combined into the composite by casting it into a liquid or pasty matrix of the other basic material which then consolidates around it to form the composite. At the other end of the range are materials, such as eutectic composition solid solutions, in which the composite material forms by simultaneous precipitation of the basic materials as randomly-distributed, microscopic precipitates, intimately mixed and usually firmly bonded within the body of the composite. In the latter case the materials are more likely to be called *complex materials* rather than composite materials but the underlying principles of behaviour are the same and the difference is mainly one of scale.

Three types of integrated composite material are distinguishable

(1) fibre reinforced materials
(2) filled materials
(3) structurally integrated materials.

6.15 FIBRE-REINFORCED MATERIALS

It is possible to produce certain materials in a form in which a large proportion of the ideal strength can be usefully developed because the material is kept relatively free of defects. Unfortunately this form is usually that of a thin filament or *fibre* which may have astonishing axial tensile strength and great flexibility but is incapable of sustaining any significant axial compressive load. The problem is to utilise this potentially high strength. The method used is to stiffen suitably disposed fibres without affecting their properties by embedding them in a continuous stiffening matrix of a suitable normal material (Plate 4).

Thus a brittle weak material, such as glass, can be drawn in a heated plastic state into long thin fibres which have a relatively orderly defect-free structure when cooled to room temperature and can have a tensile strength of about 3·0 kN/mm^2 (normal soft mild steel has a strength of about 0·45 kN/mm^2). These fibres are very flexible and can be woven into fairly strong cloth. Glass fibres intended for reinforcement are laid either in an orderly manner relative to each other, to give controlled directional strength, or in a random fashion, to give relatively isotropic lower strength, and then the arrangement is stiffened by investing it with a matrix of a suitably stiff polymer. The result is a *glass-fibre-reinforced-plastic* often loosely called *fibre-glass* although the latter term strictly applies only to the prepared glass.

The polymeric material is prepared in a liquid form which will harden after it is mixed with a suitable hardener. The liquid mixture can be prepared and mixed with the glass fibre in successive small bulk quantities which can be added to each other in place or it can be brushed or sprayed on to successive layers of fibre as they are laid down. The method depends on what is being done with the material and the hardening action time has to be adjusted to prevent premature hardening. *Chopped strand* (Plate 4a) short lengths of randomly-orientated fibres mix readily in bulk but mats of fibres (Plate 4b), loosely or finely woven into suitable sheets, respond best to brushing on. Good adhesion between the fibres and the matrix is essential so that stress can be transmitted from one to the other. It is also essential that the matrix be sufficiently elastic to take up the strain developed under load in the fibre without disintegrating. This usually means, since the matrix is likely to be much weaker than the fibre, that the former must have such a proportionately low modulus of elasticity that its elastic limit is not exceeded when it is strained with the fibre. Stress distribution in such a material may be calculated with fair accuracy if the geometry of the material, the average distribution of matrix and fibre throughout the composite, the relevant mechanical properties of the constituent materials and the applied load are known.

Glass is somewhat limited in its potential uses but other materials can be used for reinforcement in this way. Use has been made of fine-size hard-drawn (strain-hardened) metal wire and of carbon fibre with varying degrees of success. These materials are more suitable for higher-stress higher-temperature service with metal or ceramic refractory matrices. Carbon might not seem to be a very promising fibre material but the covalent tetrahedral diamond structure of carbon is one of the strongest known and a similar crystalline structure can be developed in carbon fibre produced from an organic filament by a complex system of heat treatment. Carbon fibre can have a tensile strength of perhaps 5 kN/mm^2 with an elastic modulus of about 76 MN/mm^2 (over three times that of steel). (Note that it is important that much higher elastic moduli be developed with very high strength materials if the elastic deflections under reasonable working loads are to be kept within acceptable limits.)

Another potential source of reinforcing material is found in 'whisker' growth of suitable crystalline materials, including some metals. A whisker is a thin, uniformly-sectioned single crystal which can be almost free of defects and can be grown in various ways from suitable nuclei. Growth from a suitable supersaturated liquid solution is possible, but controlled growth from the vapour state is probably more commonly used. The process is difficult to control and whisker size is likely to be very limited, the diameter being only a few thousandths of a millimetre (fibres can be larger in section, perhaps up to ten times these diameters) and, at the most, only a few centimetres long. Defects cannot always be avoided, but if better process control and bulk production should become possible and longer whiskers can be grown, there may be great possibilities in their use. As an indication iron whiskers with tensile strengths as high as 94 kN/mm^2 have been grown. There are difficulties in finding suitable matrices for metal whiskers, because high temperatures are likely to be needed when forming the material and diffusion of atoms between matrix and whiskers becomes a strong possibility both in preparation and in service.

The primary purpose of fibre composites is a development of strength but it is probable that materials of this kind will always be liable to fatigue and/or impact weaknesses because of the wide differences between the strain behaviour of the constituent materials This aspect may reduce the number of useful applications even if suitably-strong fibres can be developed.

6.16 FILLED COMPOSITE MATERIALS

Many network-structured, polymeric-structured and open crystalline-structured materials have structures which are both inherently weak and liable to attack by interpenetration of alien substances. Such a material may be strengthened and rendered less liable to penetrative attack by making it into a composite and deliberately filling its interstitial spaces with a penetrating material which will stiffen up its structure (rather like air or water filling can stiffen up a balloon), blank off the interstices and limit corrosion penetration. A penetrant of this kind, whether incorporated during the initial manufacture of the composite, or absorbed at a later stage, must itself be non corrosive to the parent structure and, preferably, should bond firmly to it to give added strength. Such an additional material is usually called a *filler*.

Alternatively, it may be possible to economise in the production of a relatively expensive parent material by diluting its volume, at some stage of manufacture, by the addition of strong bulky particles of a cheaper substitute to which the parent substance will adhere and which will give properties, principally mechanical properties, similar to those of the parent material without losing too many of the other characteristics provided by the parent material.

Another possibility is that particles of some suitable material may be required to develop their own unique properties against the background of a completely different matrix in which they are embedded. Such particles may, according to their purpose, make up only a small part or anything up to a large proportion of the whole volume of the composite.

Any one of these preceding types may constitute a *filled composite material* (Plate 5).

Network gels can sometimes provide matrices for useful filled network structures and one has already been mentioned, namely, Portland cement. Many polymers have loose, open structures which, by the incorporation of appropriate different fillers, can be changed from weak materials either into stronger, rigid, brittle materials or into notably elastic and plastic materials (a filler suitable for the latter purpose usually being called a *plasticiser*). Probably, the most widely varied uses for fillers are found in polymeric materials since fillers can be used in them not only for adjusting mechanical properties and for resisting absorption damage but also for purposes such as colouring, texture adjustment, improving electrical properties and improving fire resistance.

Concrete is a well known example of bulk filling in which a cheap aggregate of sand or sand and gravel is used to give bulk to a Portland cement mixture without detracting from the final strength of the cement. There is a limit to what can be done in this way and the cement-sand-gravel ratio is limited to a minimum ratio at which the liquid cement-water mixture can just fill all the interstices left between the particles of gravel and the particles of sand, after the constituents are properly mixed. (Note that with proper filling, the factor that governs the solid strength is then the strength of the hydrated cement which itself depends on the initial cement-water ratio.) Bulk filling is also used in clay-brick and clay-refractory manufacture when suitable old material (*grog*) is crushed and mixed in with new clay before moulding and firing.

The third type of filled material is found most commonly in cutting materials for machining. For example, diamond or other abrasive particles are suspended in a suitable rubber or ceramic base to make slitting and grinding wheels. In another cutting application tungsten carbide particles are embedded in cobalt to make *cemented carbide* (see Plate 5b) cutting tips which will cut very hard materials at quite high cutting temperatures. Another example of the same group of composites in another field of use is found in white-metal bearing materials in which a number of hard small particles of a material such as an antimony-tin compound, are used to improve wear resistance and lubrication by being embedded in the matrix of a softer bearing metal such as tin (see Plate 5).

Metallic alloys may often be adapted to make filled composites by making use of powder metallurgical methods, in which powdered particles of the required constituents are carefully mixed in appropriate sizes and proportions, compacted under pressure, and then sintered into an integrated whole by suitable interdiffusion means usually with little or no melting. A few filled metallic composites may be made by normal melting and casting methods.

6.17 STRUCTURALLY INTEGRATED MATERIALS

The structural features of most of the composite materials coming within the two groups, fibre-reinforced and bulk-filled composites, are characterised by the use of constituent materials with distinctive, mainly independently controlled, bulky particle shapes and sizes. However, in the fibre-reinforced group it is to be noted that some very small diameter, very short fibres are used in certain materials and in the second group some materials use very finely divided, almost randomly-shaped particles. These materials tend to merge into the third group the *structurally integrated composite materials* in which the constituent substances are present in microscopically-finely divided, but clearly distinguishable, phases which do not have preformed or finite structural sizes and shapes. These phases are precipitated or developed within the composite material usually in a seemingly random way, as a result of the manufacturing process.

In some cases, materials of this third group which happen to have their constituents distributed in fairly large individual phase-groups, may be difficult to distinguish from some materials in the other two type groups which have very fine subdivision of their constituents. As far as metallic materials in this group are concerned, these generally come within the scope of alloy systems since alloying is the normal mechanism for preparing them. In fact, all alloys with clearly optically-distinguishable, differing phases present within their structures are integrated composites (see Plates 1 and 2a). There are two features that separate structurally integrated composites from other composites

(1) The distinctive phases within the material are developed from differing constituents reacting within the material to produce a specific compound or a modified structure, such as a solid solution, in association with a parent phase or phases.
(2) The phases are strongly and firmly bonded to each other so that the material is a firmly integrated whole.

Because a composite material of this kind is much more uniformly structured than other composites it behaves as a whole more like a completely individual material. If directional properties are not deliberatedly created in such a material it will behave in a relatively isotropic manner relative to most, if not all, of its properties. The properties will tend to be a proportionate compromise between the individual properties and proportions of the separate phases. Thus, an inherently soft phase may be stiffened up and strengthened by being enmeshed in fine dispersion within a stiff, brittle material while the latter is made more plastic and less brittle by having its structure skeletonised around the softer enmeshed phase. Pearlite in steel is a good example of these effects (see Plate 2a), the softness of the ferrite is stiffened and the brittleness of the cementite is modified by the association.

In some materials use may be made of special thermal and/or mechanical treatments to develop directional properties in the structure by controlled directional distribution of the component phases. A notable example of this is the rapid, controlled solidification, within a strong magnetic field, of certain hard magnetic alloys which has the effect of breaking the magnetic domains down into very narrow bands lying in the direction of the applied field, subsequently leaving the material with increased permeability, higher remanence, and greater hysteresis in the direction of the applied solidification field.

There are other possibilities with materials of this type but these are too specialised to consider in the present text.

6.18 DEVELOPMENT OF NEW MATERIALS

It hardly needs stating that there are many materials which are either completely undeveloped or undiscovered. There are many reasons for this situation. Some materials are so rare, so costly to isolate, or so difficult to study that their potentialities are little understood while others remain undiscovered because no one has thought to search in their direction or someone has made a cursory search and missed their presence. The field of materials is so vast that, even although there may be many searchers it is likely that this state of affairs will continue for many years.

In what line is advance most likely to be made? This is an impossible question to answer with any certainty for the situation changes from year to year. It used to be claimed that polymeric materials were going to revolutionise engineering and replace conventional materials, but although tremendous strides have been made both in developing new polymers and applying them industrially, equally great developments have taken place in metals, non metallic refractories and glasses. If tonnage use of material throughout the world is considered it is probable that the relative proportions used have not greatly changed in recent times. Uses of wood and other natural organic materials such as rubber have decreased, mainly because local supplies are insufficient to meet the rate of present-day demand, or are too costly in terms of time, labour, and transport. Synthetic materials of the polymeric type have increased in use in this same field and have either taken over or have become partners with the original material in providing composite substitute materials.

In the present generation the greatest tonnage use of constructional material is probably concrete, metals come in second place, but ceramic non metallic refractories cannot be very far behind, because these materials usually require frequent replacement even if they are not used in the same large quantities for new constructions as are metals. Ceramics are being widely displaced by polymeric materials for many domestic uses.

6.19 FUTURE DEVELOPMENTS IN METALS

It is unlikely that a new, previously unknown, metal will be discovered, but it is certain that many new alloy combinations and methods of treatment remain to be discovered. These new alloys may have physical and chemical properties undreamed of today; they may develop superconductivity at higher temperatures (very low temperatures are needed at present), or superplasticity (limited at present to a few alloys), or universal corrosion resistance (no known metal or alloy has full resistance to every form of corrosive attack), or ideal strength or other properties not currently visualised as possible.

The main advance seems likely to be in the development of more nearly ideal strength and in the development of greater strength and resistance to elevated temperature situations. However, there are limitations to the former in that the ideal basic modulus of elasticity of metals is not very high, the highest tensile modulus being found in tungsten at 414 GN/mm^2 (42×10^9 kgf/mm^2) compared with steel at about 207 GN/mm^2 (21×10^9 kgf/mm^2). This means that even the stiffest metal will give very pronounced total elastic deformation if it is used efficiently at working stresses near to its elastic limit. This effect is aggravated by the fall off in elastic modulus as the stress rises near to the ideal elastic limit (Figure 5.19). If the stiffness-to-weight ratio is important tungsten is not quite so good as steel, its ratio of E/sg being 22 GN/mm^2, while that for steel is about 26 GN/mm^2, since tungsten has more than twice the density of steel. One of the prospectively useful metals could be beryllium with $E = 310$ GN/mm^2 and $E/sg = 167$ GN/mm^2, but beryllium is a difficult metal with which to work and it is very toxic. Molybdenum also shows promise with E at 345 GN/mm^2 and E/sg at 33·9 GN/mm^2, particularly for use at elevated temperatures, but it is difficult material to refine and cast.

From the simple stiffness point of view, metals do not look as promising as materials like alumina ($E = 379$ GN/mm^2 and $E/sg =$ 95 GN/mm^2) and silicon carbide ($E = 551$ GN/mm^2 and $E/sg =$ 172 GN/mm^2), but the inherent plasticity, comparatively good shock resistance, and ease of manufacture of most metals still makes them the most promising materials of construction. It seems that for some time to come metals will remain the preferred sources of new materials and research and development will continue.

Uses of metals have been greatly advanced by alloying them together and combining them with other materials, and giving them special treatments to make stiffer, stronger metastable alloys. Precipitation hardening treatments are evidence of this and the use of *dispersion hardening* seems likely to develop. *Dispersion hardening* entails introducing into metal lattices in dispersed form compounds which normally could not be found there. Its effect is analogous to the effects of quench-hardening in steel but it is a persistent effect at higher temperature because the dispersed particles are insoluble, but that factor also makes it much more difficult to achieve the dispersion. Metal oxides are among the most effective dispersion-hardening agents and such strengthened alloys as thoriated tungsten and thoriated nickel are well known for their strength properties at elevated temperatures. Dispersion hardening can be achieved only by special means such as powder metallurgy, mixing and treating the constituents in powder form, either directly mixing in the hardening agent in its final form, or mixing in reactive constituents and subsequently causing them to react to give their reaction product in the proper dispersion. Some metals may be heated to induce diffusion of oxygen into the interior and simultaneous oxide reaction with a suitable dissolved constituent. This method is already in use and is one with considerable potential for the future as the process becomes more widely understood.

Dispersion hardening may be superimposed on other forms of strengthening and there are possibilities for developments on these lines in many alloy systems.

As the proportion of a dispersed constituent in any given material is increased its effect can go beyond that of the concept of simple alloying and the material becomes more like a true composite in which the distinctive identity of the original parent metal is lost.

It seems unlikely that dispersion strengthening mechanisms will ever be applicable in the same way to non metallic materials although there are some vaguely similar uses in polymeric materials.

6.20 FUTURE DEVELOPMENT IN CERAMIC MATERIALS

There is strong pressure from development engineers to find materials that will give good strength and good corrosion resistance at elevated temperatures. Much of the pressure is directed at refractory metals, but there remains a very strong pressure for new types of non metallic refractories such as ceramic network-structured materials particularly for less brittle types and for better resistance to thermal shock.

Much has been done to improve the quality of materials like alumina (Al_2O_3), silica (SiO_2), and zirconia (ZrO_2) which, especially in their crystallised forms, are amongst the present most effective bases of ceramic materials, particularly for refractory uses. The operating temperatures of these materials can be quite high (alumina melts at 2300 K and silica at 1940 K and zirconia at 2770 K) and they are relatively stable and unreactive at elevated temperature. Carbon has a high melting temperature (3770 K) and is also used as a refractory. Some use is also made of silicon carbide (2870 K) although this material tends to decompose at high temperatures before it reaches its melting point. None of these materials is notably strong at elevated temperature, but they are amongst the few that are readily available for high temperature work.

No outstanding new ceramic materials have been developed recently but older ones have been greatly improved mainly by use of purer materials, and by better control of composition and manufacture. However it is possible that a break-through may take place as this field of knowledge improves and as the techniques of manufacture and control of reaction in materials at very elevated temperatures becomes more sophisticated. If means for preventing the ready propagation of fracture through such materials could be devised then great advances might be made, but brittleness seems inherent in this type of material.

6.21 FUTURE DEVELOPMENT IN POLYMERIC MATERIALS

Polymeric materials suffer from two major disadvantages in relation to mechanical strength.

(1) They are too sensitive to change in temperature.
(2) Their elastic properties are complex, sensitive to the effects of strain rate and their variations may adversely influence other properties.

Future development is bound to focus on overcoming these influences if polymeric materials are to compete with other types of materials. These disadvantages arise from the nature of the bonds operating in polymeric materials. Within the polymer chains the bonds can be exceedingly strong, but between the chains bonding is not strong, the density of bonding may be low and the spacing of bonds is liable to be irregular.

Change in temperature can greatly influence the interchain bonds. A fall in temperature may so strengthen these bonds that a previously elastic and plastic material becomes brittle hard and a brittle hard material an even harder and more brittle one. Conversely a rise in temperature may loosen these same bonds, lowering the elastic properties and increasing the plasticity. In many cases rising temperature induces chemical change within the system and may cause disintegration of the structure before melting can begin. Any of these effects can cause associated changes in physical and chemical properties.

Externally-applied stresses, by their effects on the bonding within a polymeric structure, may induce changes in electrical, optical and corrosion resisting properties, many materials being particularly sensitive to stress-corrosion effects. The rate at which stress is applied and released may greatly change the elastic and plastic properties (an extreme effect of this kind is seen in 'bouncing putty', a silicone rubber, which will deform readily with time under its own weight but will deform and recover elastically if lightly impacted by bouncing it against a hard surface or will shatter if it is impacted too heavily). This last is one of the factors which tends to rule against the use of polymeric materials for load-bearing applications in structures, since it makes it hard to control or even predict the deflection behaviour of a structure in differing service situations.

On the other hand, many polymeric substances are readily adaptable to suit special demands. By adjustment of material structural coordination and arrangement, materials ideally suited to such widely divergent applications as glue bonding, non stick sliding surfaces, infinitely plastic seals, elastic seals, rigid seals, thermal insulation, electrical insulation, high corrosion resistance, abrasion resistant paints, etc., can be developed.

High strength bonds exist within and between certain polymers and in theory it should be possible to make polymeric substances very strong. Unfortunately, the problem is to get a sufficiently high density of bonding distributed in the right direction to give a uniformly high coordination of micro-structural mechanical properties and a really high elastic limit in conjunction with a high enough elastic rate on at least one controllable axis. Crystallisation, see Figure 5.11b, offers possibilities, but so far the ideal method for combining and interconnecting polymeric chains has not been discovered. Polymers that would give good cross-linking are not suitably shaped and those that are suitably shaped do not give strong cross-linking. It may be possible to develop very long, very strong, straight polymeric chains that can be lined up alongside each other and used rather like aligned strong fibres, but at present the nearest to this is found in nylon cords and ropes, which, although much better than natural fibre cords and ropes, are still not competitive with either wire ropes or other forms of moderately strong metals.

6.22 CONCLUSIONS ON FUTURE DEVELOPMENT

It seems unlikely that any revolutionary new elemental material will be discovered but it may well be that revolutionary new combinations of existing materials may become possible either in the form of integrated composites or in the fibre-reinforced type of structure. It is also possible that a completely new method may be found for strengthening metallic alloys or other types of material.

Cermets, intimate mixtures of ceramic and metallic materials, have been the subject of much investigation, more particularly for higher temperature service, but beyond the stage of effective dispersion-hardening of metals it is doubtful if very much will be gained in this field. Brittleness and sensitivity to thermal shock appear to be inherent in nearly everything that has been tried so far in this line.

Fibre reinforcement is likely to develop further as new forms of strong fibre are prepared and better materials and new means are found for backing them up; but full use will not be made of the potential strengths of these materials, unless the problem of fatigue and similar causes of relatively unpredictable failure can be overcome. One possibility that is under investigation is the treatment of eutectics and eutectoids under such controlled conditions that one or both constituents precipitate in an ideally-strong structural form so geometrically distributed and shaped that the equivalent of a very finely structured continuous-fibre-reinforced material is achieved. The latter study is not likely to solve the problem of the relatively low values of moduli of elasticity found in most materials.

An interesting possibility in relation to strength would be the discovery of a means for crystallising a material into a lattice of similar density of packing to that of a close-packed metal, but in this case, held together almost exclusively by high strength ionic bonds instead of lower-strength metallic bonds. The stronger form of bond and the higher density of strong bonding would combine to give a high strength associated with much-higher-than-normal modulus of elasticity; whilst, perhaps, still being able to keep sufficient of the plasticity of a metal to reduce the easy initiation and propagation of fracture. One must add however that what is known about atomic bonding theory does not give even a glimmer of hope that the 'best of both worlds' answer is even a faint possibility.

Materials such as concrete have little possibility left for much change since it is their inherent cheapness that explains their widespread use. Natural material (sand and gravel) that needs little or no preparation, forms the main bulk of concrete and sets the maximum limit on strength. Some improvement might be made in the bonding cement itself, but before any marked overall improvement becomes possible the normal type of aggregate would have to be radically transformed, probably at appreciable cost, or else a completely new form would have to be discovered.

Knowledge of semiconductors is increasing and better technologies for their manufacture are being developed. It seems likely that new materials and new applications are certain to be found and will cause many changes in electronics. In particular it seems likely that, with advance in solid state circuit construction, electronically controlled automation will be introduced into more and more commonplace applications for improving the standards of living.

Resistance to strongly-corrosive environments, at temperatures greater than 700 K, is likely to be one of the most promising fields of development of polymeric materials. However, it may be that a completely new system of surface protection of other materials by the incorporation of suitable polymeric substances into their outer surface layers may become one of the means of using the good chemical properties of polymers in conjunction with the properties of other materials.

The development of tougher flexible polymers with improved electrical insulation properties for making possible the use of thinner, stronger coverings on power-conductor wires is a likely event which might greatly improve the efficiency and safety of electric power distribution.

No one can forecast with any certainty what new developments will be seen, but it can be said with certainty that the science of materials is going to continue to be one of the most important, exciting and interesting of the sciences.

BIBLIOGRAPHY

GORDON, J. E., *The New Science of Strong Materials*, Parts 2 and 3, Pelican (1968)

TWEEDDALE, J. G., *Metallurgical Principles for Engineers*, Chaps 4 and 6–13, Iliffe, London (1962)

VAN VLACK, L. H., *Elements of Materials Science*, Chaps 10–13, Addison-Wesley (1968)

YARSLEY, V. E., and COUZENS, E. G., *Plastics in the Modern World*, Penguin, Hammondsworth (1969)

Table 1 ELEMENTS ARRANGED IN ALPHABETICAL ORDER OF THEIR CHEMICAL SYMBOL

Symbol	*Name*	*Atomic number*	*Atomic mass*	*Symbol*	*Name*	*Atomic number*	*Atomic mass*
A or Ar	Argon	18	39·944	Mn	Manganese	25	54·94
Ac	Actinium	89	227	Mo	Molybdenum	42	95·95
Ag	Silver	47	107·880	N	Nitrogen	7	14·008
Al	Aluminium	13	26·98	Na	Sodium	11	22·991
Am	Americium	95	(243)	Nb	Niobium	41	92.91
As	Arsenic	33	74·91	Nd	Neodymium	60	144·27
At	Astatine	85	(210)	Ne	Neon	10	20·183
Au	Gold	79	197·0	Ni	Nickel	28	58·71
B	Boron	5	10·82	Np	Neptunium	93	(237)
Ba	Barium	56	137·36	O	Oxygen	8	16
Be	Beryllium	4	9·013	Os	Osmium	76	190.2
Bi	Bismuth	83	209·00	P	Phosphorus	15	30·975
Bk	Berkelium	97	(249)*	Pa	Protoactinium	91	(231)
Br	Bromine	35	79·916	Pb	Lead	82	207·21
C	Carbon	6	12·011	Pd	Palladium	46	106·4
Ca	Calcium	20	40·08	Pm	Promethium	61	(147)*
Cd	Cadmium	48	112·41	Po	Polonium	84	(210)*
Ce	Cerium	58	140·13	Pr	Praseodymium	59	140·92
Cf	Californium	98	(251)*	Pt	Platinum	78	195·09
Cl	Chlorine	17	35·457	Pu	Plutonium	94	(242)
Cm	Curium	96	(247)	Ra	Radium	88	(226)
Co	Cobalt	27	58·94	Rb	Rubidium	37	85·48
Cr	Chromium	24	52·01	Re	Rhenium	75	186·22
Cs	Caesium	55	132·91	Rh	Rhodium	45	102·91
Cu	Copper	29	63·54	Rn	Radon	86	(222)
Dy	Dysprosium	66	162·51	Ru	Ruthenium	44	101·1
Er	Erbium	68	167·27	S	Sulphur	16	32·006 ± 0·003
Es	Einsteinium	99	(254)				
Eu	Europium	63	152·0	Sb	Antimony	51	121·76
F	Fluorine	9	19·00	Sc	Scandium	21	44·96
Fe	Iron	26	55·85	Se	Selenium	34	78·96
Fm	Fermium	100	(253)	Si	Silicon	14	28·09
Fr	Francium	87	(223)	Sm	Samarium	62	150·35
Ga	Gallium	31	69·72	Sn	Tin	50	118·70
Gd	Gadolinium	64	157·26	Sr	Strontium	38	87·63
Ge	Germanium	32	72·60	Ta	Tantalum	73	180·95
H	Hydrogen	1	1·0080	Tb	Terbium	65	158·93
He	Helium	2	4·003	Tc	Technetium	43	(99)*
Hf	Hafnium	72	178·50	Te	Tellurium	52	127·61
Hg	Mercury	80	200·61	Th	Thorium	90	232·05
Ho	Holmium	67	164·94	Ti	Titanium	22	47·90
I	Iodine	53	126·91	Tl	Thallium	81	204·39
In	Indium	49	114·82	Tm	Thulium	69	168·94
Ir	Iridium	77	192·2	U	Uranium	92	238·07
K	Potassium	19	39·100	V	Vanadium	23	50·95
Kr	Krypton	36	83·80	W	Tungsten	74	183·86
La	Lanthanum	57	138·92	Xe	Xenon	54	131·30
Li	Lithium	3	6·940	Y	Yttrium	39	88·92
Lu	Lutetium	71	174·99	Yb	Ytterbium	70	173·04
Md	Mendelevium	101	(256)	Zn	Zinc	30	65·38
Mg	Magnesium	12	24·32	Zr	Zirconium	40	91·22

The values given normally indicate the mean atomic mass of the mixture of isotopes found in nature. Particular attention is drawn to the value for sulphur, where the deviation shown is due to variation in relative concentration of isotopes.

Bracketed values refer to the individual isotopes of radioactive elements. In most cases the value for the most long lived is given. Where, however, an asterisk occurs the atomic mass is that of the better known isotope.

Table 2 ELECTRON STRUCTURE OF THE ELEMENTS UP TO RADIUM

		Quantum Number																	
	Principal	1	2		3			4				5				6			7
	Secondary	s	s	p	s	p	d	s	p	d	f	s	p	d	f	s	p	d	s
	Quantum Capacity For Electrons	2	2	6	2	6	10	2	6	10	14	2	6	10	14	2	6	10	2
Atomic No.	*Element*																		
1	H	1																	
2	He	2																	
3	Li	2	1																
4	Be	2	2																
5	B	2	2	1															
6	C	2	2	2															
7	N	2	2	3															
8	O	2	2	4															
9	F	2	2	5															
10	Ne	2	2	6															
11	Na				1														
12	Mg				2														
13	Al				2	1													
14	Si	Filled			2	2													
15	P				2	3													
16	S				2	4													
17	Cl				2	5													
18	A				2	6													
19	K							1											
20	Ca							2											
21	Sc						1	2											
22	Ti						2	2											
23	V	Filled					3	2				Transition Elements (Manganides)							
24	Cr						5	1											
25	Mn						5	2											
26	Fe						6	2											
27	Co						7	2											
28	Ni						8	2											
29	Cu							1											
30	Zn							2											
31	Ga							2	1										
32	Ge	Filled						2	2										
33	As							2	3										
34	Se							2	4										
35	Br							2	5										
36	Kr							2	6										
37	Rb										Empty	1							
38	Sr											2							
39	T									1		2							
40	Zr									2		2							
41	Nb									4		1				Transition Elements (Technetides)			
42	Mo	Filled								5		1							
43	Tc									6		1							
44	Ru									7		1							
45	Rh									8		1							
46	Pd									10									
47	Ag											1							
48	Cd											2							
49	In											2	1						
50	Sn	Filled										2	2						
51	Sb											2	3						
52	Te											2	4						
53	I											2	5						
54	Xe											2	6						
55	Cs											2	6		Empty	1			
56	Ba											2	6			2			
57	La											2	6	1		2			
58	Ce										2	2	6			2			
59	Pr										3	2	6			2			
60	Nd										4	2	6			2			
61	Pm										5	2	6			2			
62	Sm										6	2	6			2			
63	Eu	Filled									7	2	6			2			
64	Gd										7	2	6	1		2			
65	Tb										9	2	6			2			
66	Dy										10	2	6			2			
67	Ho										11	2	6			2			
68	Er										12	2	6			2			
69	Tm										13	2	6			2			
70	Yb										14	2	6			2			
71	Lu										14	2	6	1		2			
72	Hf													2		2			
73	Ta													3		2			
74	W													4		2	Transition Elements (Rhenides)		
75	Re	Filled												5		2			
76	Os													6		2			
77	Ir													7		2			
78	Pt													8		2			
79	Au															1			
80	Hg															2			
81	Tl															2	1		
82	Pb	Filled														2	2		
83	Bi														Empty	2	3		
84	Po															2	4		
85	At															2	5		
86	Rn															2	6		
87	Fr	Filled														2	6		1
88	Ra															2	6		2

Table 3

PERIODIC TABLE OF THE ELEMENTS, IN ORDER OF ATOMIC NUMBER AND GIVING PERIOD NUMBERS AND NAMES OF SOME SUBGROUPS

IA	*IIA*							1 H				*IIIA*	*IVA*	*VA*	*VIA*	*VIIA*	2 He
3 Li	4 Be											5 B	6 C	7 N	8 O	9 F	10 Ne
11 Na	12 Mg	*IIIB*	*IVB*	*VB*	*VIB*	*VIIB*	*VIII*			*IB*	*IIB*	13 Al	14 Si	15 P	16 S	17 Cl	18 A
19 K	20 Ca	21 Sc	22 Ti	23 V	24 Cr	25 Mn	26 Fe	27 Co	28 Ni	29 Cu	30 Zn	31 Ga	32 Ge	33 As	34 Se	35 Br	36 Kr
37 Rb	38 Sr	39 Y	40 Zr	41 Nb	42 Mo	43 Tc	44 Ru	45 Rh	46 Pd	47 Ag	48 Cd	49 In	50 Sn	51 Sb	52 Te	53 I	54 Xe
55 Cs	56 Ba	57-71 RARE EARTHS	72 Hf	73 Ta	74 W	75 Re	76 Os	77 Ir	78 Pt	79 Au	80 Hg	81 Tl	82 Pb	83 Bi	84 Po	85 At	86 Rn
87 Fr	88 Ra	89- ACTINIDES		(BCC)	(BCC)			(FCC)	(FCC)	(FCC)	(HCP)				CHALCOGENS	HALOGENS	RARE GASES
ALKALI METALS (BCC)	ALKALI EARTHS (HCP)																

57 La	58 Ce	59 Pr	60 Nd	61 Pm	62 Sm	63 Eu	64 Gd	65 Tb	66 Dy	67 Ho	68 Er	69 Tm	70 Yb	71 Lu	RARE EARTHS (LANTHANIDE SERIES)
89 Ac	90 Th	91 Pa	92 U	93 Np	94 Pu	95 Am	96 Cm	97 Bk	98 Cf	99 E	100 Fm	101 Mv	ACTINIDE SERIES		

Table 4 PROPERTIES OF SOME ELEMENTAL SOLIDS

Material	*Symbol*	*Atomic number*	*Normal valency*	*Atomic mass*	*Effective atomic Diameter pm*	*Normal structure*	*Specific gravity*	*Melting point K*	*Boiling point K*	*Resistivity* $\Omega m \times 10^{-6}$	*Thermal conductivity* $W/m°C$	*Normal tensile strength* N/mm^2	*Normal tensile modulus* kN/mm^2
Aluminium	Al	13	3+	26·97	286	FCC	2·699	933	2333	265	230·0	80	69
Antimony	Sb	51	5+	121·76	29	Rhomb.	6·62	903	1713	3900	18·8	100?	78
Arsenic	As	33	3+ 5+	74·91	25	Rhomb.*	5·73	373V	?	3500	?	?	?
Barium	Ba	56	2+	137·36	434	BCC*	3·5	977	1913	?	?	?	?
Beryllium	Be	4	2+	9·02	222	CPH*	1·82	1550	3043	590	159·0	620	255
Bismuth	Bi	83		209·00	311	Rhomb.	9·8	544	1793	10680	8·4	?	32
Boron	B	5	3+	10·82	174?	Orth.	2·3	2570	?	0.18×10^{15}	?	?	?
Cadmium	Cd	48	2+	112·41	297	CPH	1·55	594	1038	680	92·0	?	55
Carbon	C	6	4+	12·01	142	Hex.*	2·22	3970	5100	137×10^3	170·0?	?	685?
Chromium	Cr	24	3+T	52·01	249	BCC*	7·19	2160	2773	1300	96·0?	290?	290
Cobalt	Co	27	2+T	58·94	250	CPH*	8·9	1768	3173	625	69·0	690	207
Copper	Cu	29	1+	63·54	255	FCC	8·96	1356	2254	170	390·0	215	~110
Gallium	Ga	31	3+	67·92	244	Orth.*	5·91	303	2343	5300	?	?	?
Germanium	Ge	32	4+	72·60	245	Diamnd cubic	5·36	1230	2033	8.9×10^6	?	?	?
Gold	Au	79	1+	197·20	288	FCC	19·32	1336	3243	219	290·0	85	837
Iridium	Ir	77	4+T	193·10	271	FCC	22·5	2727	5570	530	58·5	480?	517
Iron	Fe	26	2+ 3+ T	55·85	248	BCC	7·87	1812	3010	970	75·0	360	196

Lead	Pb	82	2+ 4+	207·21	349	FCC	11·34	600	894	2065	34·7	14	180
Magnesium	Mg	12	2+	24·32	319	CPH	1·74	923	1383	446	159·0	100?	45
Manganese	Mn	25	2+T	54·93	224	complex C*	7·43	1518	2423	18500	?	900?	159
Molybdenum	Mo	42	4+T	95·95	272	BCC	10·2	2890	5070	520	146·0?	580	328?
Nickel	Ni	28	2+	58·69	249	FCC	8·9	1728	3003	680	92·2	480	207
Niobium	Nb	41	5+T	92·91	285	BCC	8·57	2688	?	1310	52·4	275	345?
Platinum	Pt	78	3+T	195·23	277	FCC	21·45	2046	4683	980	71·2	120	145
Rhodium	Rh	45	3+T	102·91	268	FCC	12·44	2239	4773	450	88·0	300?	290
Selenium	Se	34	2—	78·96	232	Hex.*	4·81	493	953	?	?	?	?
Silicon	Si	14	4+	28·06	235	Diamnd cubic	2·33	1703	2573	10×10^6	83·8	?	113
Silver	Ag	47	1+	107·88	288	FCC	10·49	1234	4283	160	420·0	110	76
Sodium	Na	11	1+	23·00	371	BCC	0·97	371	1165	420	134·0	?	?
Sulphur	S	16	2— 6+	32·07	212	Orth.*	2·07	392	718	200×10^{21}	0·03	?	?
Thorium	Th	90	4+	232·12	359	FCC	11·5	2073	?	1900	?	300?	?
Tin	Sn	50	4+	118·70	302	BC Tetr.*	7·3	505	2543	1150	67·0	36	41
Titanium	Ti	22	4+T	47·90	291	CPH*	4·54	1933	?	8000	95·3	386	104
Tungsten	W	74	4+T	183·90	273	BCC*	19·3	3683	6200	550	170·0?	600	400
Vanadium	V	23	3+ 5+ T	50·95	263	BCC	6·0	2008	3673	2650	32·0	480	128
Zinc	Zn	30	2+	65·38	266	CPH	7·18	692	1179	590	113·0	140	97
Zirconium	Zr	40	4+T	91·22	316	CPH*	6·5	2130?	?	4100	17·0	250	82

T = Transition element. V = Begins to vaporise. * = Other structural states are known and possible.
? = Value given either a rough approximation or not unique. If no value given either irrelevant, not known or not determinable.

Table 5 SOME CHEMICAL COMPOUNDS AND THEIR USES
(g = gas, l = liquid, s = solid)

Material	*Formula*	*Normal state*	*Principal use*
Acetylene	C_2H_4	g	Raw material for some polymers
Acrylic acid	$CH_2:CH\cdot COOH$	l	Base for acrylic resins
Alabaster	$CaSO_4\cdot 2H_2O$	s	Plaster
Alumina	Al_2O_3	s	Refractory
Aluminium nitride	AlN	s	Hard surfacing steels
Aniline	$C_6H_5NH_2$	l	Base for some polymers
Barium oxide	BaO	s	Manufacture of glass
Benzene	C_6H_6	l	Base of many organic materials
Boron carbide	B_4C	s	Abrasive and wear resistant
Butane	C_4H_{10}	g	Base of some synthetic rubbers
Calcium carbonate	$CaCO_3$	s	Lime and cement
Calcium hydroxide	$Ca(OH)_2$	s	Cements
Calcium oxide	CaO	s	Cements
Calcium silicate	$CaSiO_3$	s	Glass and cements
Calcium sulphide	CaS	s	Luminous paints
Carbon disulphide	CS_2	l	Rayon manuf.
Cellulose	$(C_6H_{10}O_5)_n$	s	Base of paper, rayon, wood, etc.
Cementite	*see* Iron carbide		
Chloroprene	$CH_2CH:CCl\cdot CH_2$	l	Some synthetic rubbers
Cupric oxide	CuO	s	Glass making
Cuprous oxide	Cu_2O	s	Glass making
Cyclohexane	C_6H_{12}	l	Making some plastics
Deuterium oxide	D_2O	l	Moderator for some atomic reactions
Diaminohexane	$H_2N(CH_2)_6NH_2$	s	Nylon manuf.
Dichloroethane	$C_2H_4Cl_2$	l	PVC manuf.
Ethyl carbamate	$NH_2COOC_2H_5$	s	Urethane resins manuf.
Ethylene	C_2H_4	g	Polythene manuf.
Fluorine	$C_{13}H_{10}$	s	Synthetic resin manuf.
Formaldehyde	$HCHO$	g	Some plastics manuf.
Furfuraldehyde	C_4H_3OCHO	l	Synthetic resin manuf.

Table 5 (*contd.*)

Material	*Formula*	*Normal state*	*Principal use*
Glycerol	$CH_2OH \cdot CHOH \cdot CH_2OH$	l	Explosives and plastics manuf.
Glyoxal	$(CHO)_2$	s	Plastics manuf.
Guanidine	$HN{:}C(NH_2)_2$	s	Explosives and plastics
Hexamine	$(CH_2)_6N_4$	s	Vulcanised rubbers manuf.
Hydrazine	$H_2N \cdot NH_2$	l	Rocket propellant
Iron carbide	Fe_3C	s	Constituent of steels
Isoprene	$CH_2{:}CH \cdot C(CH_3){:}CH_2$	l	Base of natural rubber
Litharge	PbO	s	Glass, glazes and paints
Magnesia	MgO	s	Refractory bricks
Magnesium chloride	$MgCl_2$	s	A cement
Melamine	$C_3H_6N_6$	s	Some synthetic resins
Methacrylic acid	$CH_2{:}C(CH_3)COOH$	l	Plastics
Oxydiacetic acid	$O(CH_2COOH)_2$	s	Some plastics
Pentaerythritol	$C(CH_2OH)_4$	s	Some plastics
Phenylethylene	$C_6H_5 \cdot CH{:}CH_2$	l	Base of certain plastics
Sebacic acid	$HOOC(CH_2)_8COOH$	s	Some plastics
Silica	SiO_2	s	Base of some glasses and a refractory
Silicon carbide	SiC	s	High temperature abrasive
Sodium silicate	Na_2SiO_3	s	Paper and cement
Tetrafluoroethylene	$CF_2{:}CF_2$	g	Base of a plastic material
Thorium dioxide	ThO_2	s	A refractory
Titanium dioxide	TiO_2	s	A pigment in plastics and refractories
Tungsten carbide	WC	s	Abrasive
Urea	$CO(NH_2)_2$	s	Base of some synthetic resins
Vinyl acetate	$CH_2{:}CHOOCCH_3$	l	Base of a plastic
Vinyl chloride	$CH_2{:}CHCl$	g	Base of a plastic
Water	H_2O	l	Varied
Zinc chloride	$ZnCl_2$	s	Wood preservation
Zinc oxide	ZnO	s	Glass manuf.
Zinc sulphate	$ZnSO_4 \cdot 7H_2O$	s	Paper manuf.
Zirconium dioxide	ZrO_2	s	A pigment and a refractory
Zirconium silicate	$ZrSiO_4$	s	A refractory

Plate 1. Microstructure of a eutectic of copper (white) in copper-phosphide (grey) X300 (Courtesy: D. L. Thomas)

Plate 2. Changing solubility of carbon in iron at n.t.p. in nearly stable conditions (a) Microstructure of eutectoid pearlite in plain carbon steel X500. Cementite dark, ferrite light. (Courtesy: D. L. Thomas).

Plate 2. Changing solubility of carbon in iron at n.t.p. in nearly stable conditions (b) Microstructure of a grey cast iron X500 showing graphite flakes (black)

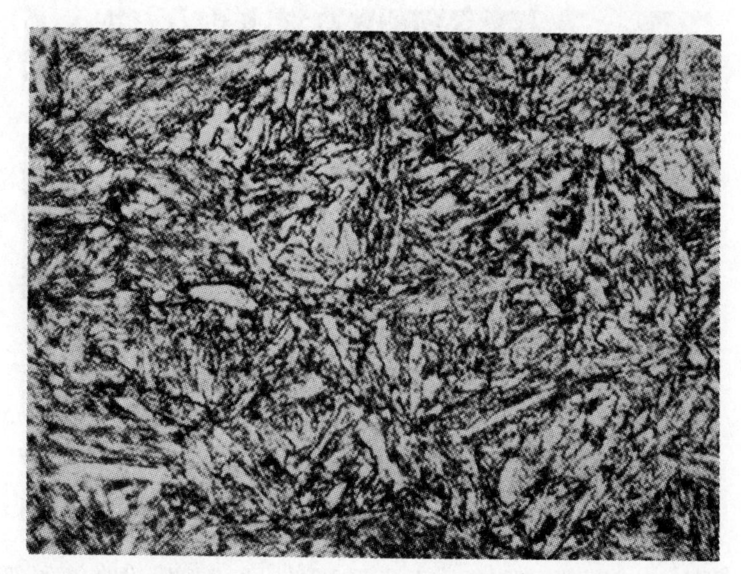

Plate 3. Lightly tempered martensite, structure beginning to resolve. Steel containing 0.8%C (×1,000). (Courtesy: D. L. Thomas)

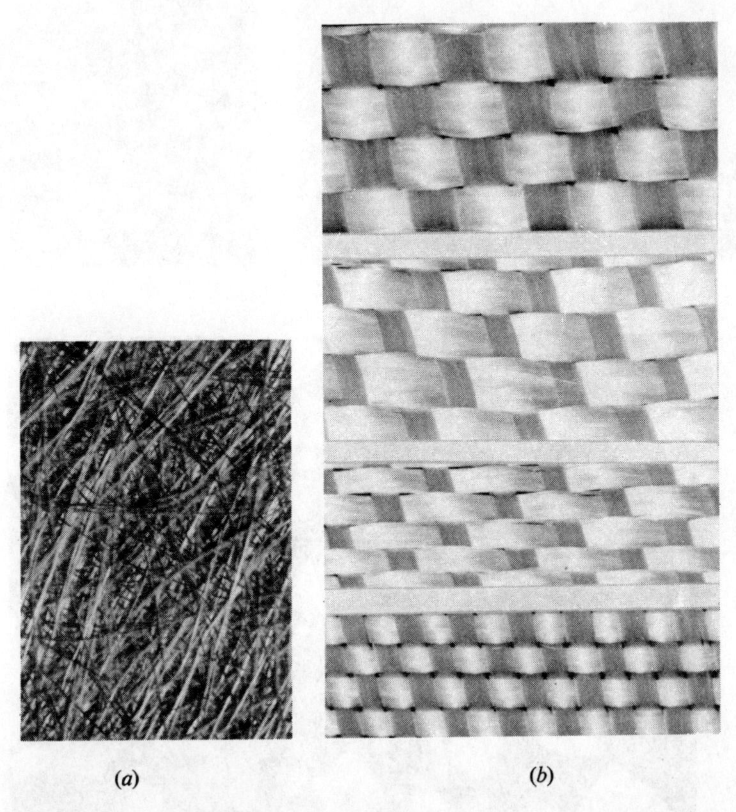

Plate 4. Forms of glass fibre for reinforcing compostic materials. (Courtesy: Fibreglass Ltd.). (a) Chopped strand random mat X5. (b) Four varieties of 'woven roving' X1.5

(*a*) (*b*) (*c*)

Plate 5. Examples of structurally integrated composite materials. (*a*) *Structure of cement mortar X10. Sand particles look like domes. Cement matrix light grey. Dark areas are granite chips.* (*Courtesy: Cement and Concrete Association*). (*b*) *Structure of cemented carbide X10 000 94% WC 6% Co. Whitish areas are Co.* (*Courtesy: T. Raine*). (*c*) *Structure of white metal bearing material X50. Sb–Sn compound particles are white and Sn with Sb traces are dark.* (*Courtesy: D.L. Thomas*)

Index

Age hardening, natural, 170
Ageing, 169
  artificial, 170
Agglomeration, 41, 67
Aggregation, 66
Allotropy, 35, 57
Alloy additions to steel, 172
Alloying, solubility changes in, 165–172
Alloys, 183
Alumina, 58, 184, 185
Aluminium, 58
Ampere, 89
Application, properties of, 7
Argon, 55
Atomic diameter, 20
Atomic mass number, 19
Atomic mass unit, 19
Atomic nature of material, 3–4
Atomic number, 16
Atomic shells, 19
Atomic weight number, 19
Atoms, 3, 12, 38, 39, 117
  behaviour of different types, 105, 106
  describing characteristics of, 13
  energy balance between, 43
  force acting on, 128
  interactions, 4
  main feature of, 14
  makeup of, 13
  mass of, 19
  types of, 13
Austenite, 171

Bainite, 171
Barrel, 84
Bearing materials, 181
Bending test, 83
Beryllium, 19, 184
Blast furnace, 153
Body-centred cubic array, 64
Boiling, 59
Boiling point, 35
Bonding agent, 66
Bonds and bonding, 4, 29, 41, 88, 90, 106, 114, 120, 123, 137, 138, 144, 157, 187, 188
  and chemical properties, 55–58
  covalent, 38, 47–51
  density of, 124
  double, 124
  electron dispersion, 38, 53
  external, 59
  heterogeneity of, 39
  hydrogen bridge, 38, 54–55
  intercrystalline, 67
  intermediate or bridge, 58
  ionic, 38, 45–47
  metallic, 38, 51–53, 65
  molecular polarisation, 38, 53
  proportionate relative orientation of, 124
  relative densities of, 123–124
  relative distributions of, 124
  types of, 38, 123
Bouncing putty, 187
Boundary defects, 115
Brass, 167
Brinell system, 77
Brittle materials, 83, 156, 175, 178, 182
Buckling, 84

Carbon, 19, 57, 185
Carbon atoms, 29, 121
Carbon fibre, 179
Cast irons, 168
Castability, 6
Catalyst, 104
Cavity defects, 116
Cemented carbide, 181
Cementite, 168, 171, 182
Ceramic materials, 122
  future development in, 185–186
Cermets, 188
Chain molecular structures, 121
Chain polymeric materials, 155, 158
Charpy V-Notch Impact Test, 76
Chemical activity, 57
Chemical compounds and their uses, table of, 195–196
Chemical properties, 3, 98, 125
  and bonding, 55–58
  ideal and real, 142
Chemical reactions, 56, 57
  gas release, 145
  indirectly involved constituents and, 145–146
  influence of reaction products, 143
  influence on own environment, 144–145
  stress effects, 145
Chopped strand, 178
Chromium, 175
Cleavage strength, 129
Close-packed-hexagonal array, 65
Close-packed plane, 64, 112
Coalescence, 169, 170
Coefficient of thermal expansion, 95, 139
Cold rolling, 158
Cold working, 154, 157
Combustion, 56
Complex materials, 177
Composite materials, 177
  filled, 179–181
  integrated, 177
  structurally integrated, 181–182
Composition, use of major changes in, 160
  use of minor changes in, 158–160
Compounds, 58
  effect of formation of, 164
Compression, 71, 73
  plasticity in, 85
Compression modulus of elasticity, 85
Compression properties, 84–85
  temperature effects, 85
Compression strength, 85
Compression yield strength, 85
Compressive force, 32
Compressive strain, 72
Compressive stress, 72
Concrete, 180, 188
Convection, 36
Cooling, 163
Cooling rate, 141
Coordinating number, 49, 52
Corrosion, 98
  atmospheric, 100
  catalytic, 104
  direct external attack, 100
  direct internal breakdown, 99
  electrochemical, 102
  electrolytic, 103
  indirect attack, 102
  irradiation-assisted, 104
  types of, 99
Corrosion resistance, 98
Corrosive environments, 189
Cost of shaping, 174–175
Coulomb, 89
Crack opening displacement, 73, 76
Crack opening displacement tests, 83
Creep, in compression, 85
  in shear loading, 87
  in tension, 83
Creep strength, 73, 75, 84
Cross-linking, 187
Crystalline materials, 40, 156
Crystalline structure, 47, 51, 127
  basic, 119
  molecular, 120
Crystalline, 187
Crystallised polymers, 121
Crystallographic planes, 61
  orientations of, 127
  reactions between, 127
Crystals, 5, 27, 40, 59–65
  molecular, 120
Curie temperature, 94
Current, 89
Cutting materials, 181

Decalescence, 141
Defects, causes of, 149
  in materials, 106
  in real materials, 133
  influence of, 117

Defects *continued*
interstitial, 107
line, 107
macrostructural, 107, 113
microstructural, 107–113
inhibition of effects of, 160
removal of, 154
by structural refinement, 150–154
structural, 106, 133
substitutional, 107
surface, 111
Deformation, elastic, effects of end restraint on, 131
plastic, 81, 82, 129, 154
Degassing, 160
Density, 139
Deoxidation of molten steel, 143
Diamagnetism, 91
Diamond, 57, 181
Dielectric, 90
Dielectric constant, 90
Diffusion, 30, 32, 68, 99, 141–143
rate of, 144, 169, 170
Directional distribution, 182
Directionality in structure, 156
Dislocation mills, 134
Dislocations, 108–111, 133, 143, 157, 160, 165
edge type, 109, 112, 133
screw type, 109
sessile, 111
stress activated, grain boundaries of, 134
Dispersion hardening, 164, 184–185
Distillation, 151
Distortion of structure, 162
Ductility, 82, 83
Duralumin, 170
Dynamite, 57

Elastic deformation, effects of end restraint on, 131
Elastic hardness, 78
Elastic limit, 81
Elastic modulus, 80, 184
Elastic rate, 80
Elasticity, 79
Electric charge, 89
Electric potential, 17
Electrical conductivity, 138
Electrical energy, 89
Electrical fields, 46, 88
Electrical power, 90
Electrical properties, 88–91
and thermal properties, 94
ideal and real, 137
Electrical units, 89
Electrolysis, 151
Electrolyte, 103
Electromagnetic force, 90
Electromagnetic properties, 32
Electron behaviour, 21, 22
Electron disposition, 25
Electron orbitals, 21, 22–27, 43
Electron pairing, 25
Electron shell, 19
Electron spin, 21
Electron transfer, 96
Electronegative zone, 21
Electrons, 19, 20, 34
free, 34
in unfilled sub-shell, 94
valency, 26, 28, 36, 43, 48–49, 51, 52, 55, 94, 97, 137
Electronvolt, 90
Electropositive zone, 21
Electro-potential field, 25, 49
Elemental solids, properties of, 193
Elements, 27
table of, 190–192
Elongation, 82–83
Endothermic reaction, 144
Endurance, 74
Endurance limit, 74
Energy, 12
interaction of specific quantities, 12
nature of, 12
types of, 12
Energy balance between atoms, 43
Energy groupings, 3, 12
Environment, effects of, 174
Eutectic reaction, 162
Eutectoid reaction, 163
Evaporation, 35
Exothermic reaction, 144

Face-centred cubic array, 64
Face-centred cubic system, 122
Fatigue, in compression, 85
in shear loading, 87
thermal, 96
Fatigue limit, 74, 75
Fatigue loading in tension, 83
Fatigue properties, 74–75
Fatigue strength, 73–75
Ferrite, 168, 171, 182

Ferromagnetism, 91
Fibre-glass, 178
Fibre-reinforced materials, 177–179, 188
Fibre structure, 157
Fibres, mechanical interlocking of, 33
Fibrous material, 177
Flow stress, 136
Fluorine, 50
Flux density, 92
Flux reaction, 151
Force acting on atoms, 128
Force interaction in ideal materials, 131
Foreign atom atmosphere, 136
Formability, 3, 6
Freezing, 35

Gamma radiation, 13
Gamma rays, 20
Gas, 29–30
  monatomic, 55
  perfect, 30
Gas Constant, 30
Gas-fixing agents, 160
Gas law, 30
Gas release in chemical reactions, 145
Gaseous state, 30
Gauss, 92
Gel structures, rigid, 123
Glass-fibre-reinforced-plastic, 178
Glass fibres, 178
Glide planes, 137
Grain boundary, 5, 115, 116
  in ideal materials, 132–133
  irregular distribution of, 133
  of stress activated dislocations, 134
Grain enlargement, 155
Grain growth, 68, 156
Grain size, 69, 155
  fine, 155–156
Grains, 5
Graphite, 57, 120
Gross impurities, 116

Half-life, 13
Hardening, dispersion, 164, 184–185
  of steel, 171
  precipitation, 169, 170
  quench, 171–172
Hardness, 73, 77
  elastic, 78
Hardness tests, 77
Heat capacity, 95
Heat energy, 94
Heat transfer, 96
Heat treatment, 156–157
Heating rate, 141
Helium, 29, 45, 55
Heterogeneous nucleation, 139–141, 150
Hexagonal-close-packed array, 65
Hexagonal-close-packed system, 112
Homogeneous nucleation, 139–141
Hook's law, 130
Hydrocarbon molecules, 50
Hydrogen, 56, 103
Hydrogen atom, 16, 29, 45, 50
Hydrogen bridge bonds, 38, 54–55
Hydrogen fluoride, 50
Hydrogen molecule, 48

Impurities, 41, 146, 150–152
  change in nature of, 159
  gross, 116
Infra-red radiation, 104
Innoculants, 150
Insulator, 90
Interstitial defect, 107
Interstitial solid solutions, 167
Interstitial solubility, 161
Iron, 19, 24, 168
Iron–carbon alloys, 168
Iron–carbon system, 167–168
Iron ore, 153
Iron sulphide, 160
Irradiation, 13, 20, 165
Irradiation-assisted corrosion, 104

Jog, 111
Joule, 89

Kelvin effect, 97
Krypton, 55

Laminated sheets, 175–176
Latent heat of change, 95
Line defects, 107
Liquid, 29–30
  pressure effects, 36–37
Liquid state, 30
Long chain molecules, 121

Machinability, 6
Machining, 181
Macrostructural characteristics, 41
Macrostructural defects, 107, 113
Macrostructure, 65
Magnesium, 29
Magnetic alloys, 182
Magnetic behaviour, forms of, 91
Magnetic domain, 91, 113–114
Magnetic fields, 91–93
Magnetic hysteresis loop, 93
Magnetic intensity, 92
Magnetic properties, 22, 91–94
  heat effects, 94
  ideal and real, 138
Magnetisation, and strain, 93
  dimensional effects of, 93–94
Magnetism, 25, 92
Malleability, 85
Manganese, 143, 160
Manganese sulphide, 160
Martensite, 171
  tempered, 171
Mass of atom, 19
Materials, atomic nature of, 3–4
  basic forms of, 175
  combined, use of, 148
  crystalline, 40
  definition, 1, 12
  factors governing characteristics of, 38
  future development, 187–188
  integrated, use of, 148
  makeup of, 38–69
  matching to application, 147
  nature of, 12–37
  nature of available, 174
  new, development of, 183
    synthesis of, 148
  properties of. *See* Properties and under specific properties
  simultaneous use of separate, 173–174
    factors controlling, 174
Materials technology, range of, 2
Matter, states of, 29–32
Mechanical properties, 32–34, 70–78
  ideal and real, 126–130, 135
  in real materials, 133
  prediction from theoretical concepts, 126
  structural basis of, 87–88
  temperature effects, 83
Melting, 35, 59, 139, 140
Mendelevium, 16
Mesons, 18
Metal oxides, 184–185
Metals, future developments in, 183
Metastability, 142
  effect of degree of, 164
Metastable state, 32
Methane, 51
Microfissures, 116, 160
Microstructural characteristics, 41
Microstructural defects, 107–113, 160
Microstructural distribution, geometry of, 159
Microstructure, change of order in, 157–158
Miller indices, 61
Misorientation of structural group, 113
Modulus of elasticity, 184
  compression, 85
Modulus of rigidity, 86
Modulus of rupture, 83
Modulus of section, 83
Molecular boundaries, 5
Molecular crystals, 120
Molecular polarisation bonding, 35, 53
Molecular structures, 29
Molecules, 5, 39, 58–59, 117
  long chain, 121
Molybdenum, 184
Monomer, 58, 121

Necking, 82
Neon, 55
Network structures, rigid, 122
Neutrons, 15, 17, 19
Nitro-glycerine, 57
Normalising, 156
Notch impact strength, 73, 75
Notch sensitivity tests, 83
Nuclear core, 43
Nuclear forces, short range, 18
Nucleation, heterogeneous, 139–141, 150
  homogeneous, 139–141
Nuclei, 35
Nucleons, 15
Nucleus, 14, 15
  solidification, 114

Octohedral planes, 64
Ohm, 90
Orbital. *See* Electron orbitals

Overhead electric power transmission, 173
Oxidation, 101
Oxides, 58
Oxygen, 56
Oxygen molecule, 49

Paramagnetism, 91
Particle size reduction, 155
Particles, 13, 15
  in electrical field, 88
  negatively charged, 89
  see also specific types of particle
Pearlite, 168, 182
Periodic table, 23, 27–28
Peritectic reaction, 163
Peritectoid reaction, 163
Permeability, 92
Petroleum, 151
Phase, 31, 182
Phase boundaries, 143
Phonons, 34
Photons, 34
Physical properties, 3, 32, 70
Plastic deformation, 81, 82, 129, 154
Plastic flow, 136
Plastic strain, 80, 81
Plasticity, 6, 79, 85, 158
Plywood, 175
Point defects, 107
Poisson's ratio, 72, 86
Polarisation, 90, 91
Polonium, 13
Polycrystalline structure, 41
Polycrystallisation, 41
Polymeric materials, 58, 183, 189
  chain, 155, 158
  crystallised, 121
  flexible, 189
  future development in, 186
  temperature effects, 186
Polymorphous substances, 57
Potential difference, 89
Powder metallurgy, 181, 185
Precipitation, 164, 169
Precipitation hardening, 169, 170
Precipitation treatment, 170
Preferred orientation, 68
Pressure, absolute zero, 36
Pressure effects, 36, 57
Principal Quantum, 28, 42, 55
Principal Quantum Numbers, 22–25, 27, 42, 44
Principal Shells, 22
Principal stresses, 72
Production scale limitations, 150
Proof stress, 81
  shearing, 86
Properties, 2
  and structure, correlation between, 123–126
  control of, 147–189
  effect of exposed surfaces, 124
  ideal, 105–106
  matching to application, 147
  of application, 7
  theoretical prediction and empirical result, 105, 106
  see also under specific properties
Protons, 15, 19
Purity, commercial, 165
  how to achieve, 150
  required standard of, 151–152
  super, 152

Quanta, 22
Quench-cracking, 172
Quench hardening, 171–172
Quenching, 169

Radiation, 13, 20, 104, 165
  and corrosion, 104
Radioactive decay, 13
Radioactive isotopes, 13, 18
Radium, 13, 25
Radon, 55
Reaction products, influence on chemical reaction, 143
Recalescence, 140
Recrystallisation, 154
Reduction of area, 82–83
Reheating, 163, 171
Reinforced concrete, 173
Relative permeability, 92–93
Remanence, 93
Resistance, 90
Resistivity, 90, 138
Rigid gel structures, 123
Rigid network structures, 122
Rockwell system, 77

Safety factor, 73
Seebeck effect, 97
Semiconductors, 90, 152, 188

Shaping, cost of, 174–175
Shear modulus of elasticity, 86
Shear strain, 72
Shear strength, 129
  testing for, 86
Shear stress, 72, 73
Shearing, 71, 73
Shearing force, 32
Shearing fracture, 87
Shearing proof stress, 86
Shearing properties, 85–87
Shearing yield stress, 86
Shock loading, 75, 87
Shore system, 78
Silica, 185
Silicon carbide, 184, 185
Silicone rubber, 187
Silver, 28
Sintering, 162
Slag, 151
Slip, 137
Sodium chloride, 46
Solid solubility, change of, 163
Solid solution, 161, 162, 166
  interstitial, 167
  substitutional, 167
Solid state, 31
Solid-state reactions, 163
Solidification, 116, 139, 154, 156
  effect of nucleation mode on, 140
Solidification nucleus, 114
Solids, 29–30, 41, 66
  elemental, properties of, 193
  pressure effects, 37
Solubility, changes in alloying, 165–172
  effect of change in, 162
  effect of degree and type of, 161
  interstitial, 161
  substitutional, 161
Solution treatment, 169
Space lattice, 59
Specific heat, 95
Stability, 34–36
Stacking fault, 112
Stainless steel, 101
Steels, 168
  alloy additions, 172
  hardening of, 171
Stepped yield, 136
Strain, 71
  and magnetisation, 93
  compressive, 72
  engineering, 79
  plastic, 80, 81
  shear, 72
  tensile, 71
  transverse contractional, 71
  true, 79
Strain hardening, 136, 156
Stress, and chemical reactions, 145
  compressive, 72
  principal, 72
  proof. *See* Proof stress
  shear, 72, 73
  tensile, 71
  yield. *See* Yield stress
Stress concentrations, 75
  critical, 136
  in ideal materials, 130
Stress-corrosion, 187
Stress–strain diagram, 79, 81, 82, 86, 136
  ideal, 130, 135
Structural defects, 106, 133
Structural refinement, 150–154
Structure, adaptation and modification of, 148, 149
  and mechanical properties, 87–88
  and properties, correlation between, 123–126
  basic crystalline, 119
  chain molecular, 121
  controllability of, 3, 4
  crystalline, 47, 51, 127
  directionality in, 156
  distortion of, 162
  fibre, 157
  heterogeneity of, 41
  molecular crystalline, 120
  rigid gel, 123
  rigid network, 122
  solid, types and characteristics of, 117–123
  synthesis of new, 148
  uniformity of, 154
Sub-grains, 113
Sublimation, 35
Substitutional defect, 107
Substitutional solid solutions, 167
Substitutional solubility, 161
Sulphur, 159–160
Super-lattice, 120
Super purity, 152
Superconductivity, 91
Superheating, 141
Supersaturation, 164, 170
Surface defects, 111

Surfaces, exposed, effect on properties, 124

Temperature effects, 83, 85, 87, 94, 186
Tempering, 171
Tensile force, 32
Tensile properties, 73, 78–84
  temperature effects, 83
Tensile strain, 71
Tensile strength, 82
Tensile stress, 71
Tension, 71
Tesla, 92
Thermal conditions, 57
Thermal conductivity, 95
Thermal cycling, 75
Thermal diffusivity, 95
Thermal energy, 34–36
  exchange of, 144
  transfer of, 96
Thermal expansion, coefficient of, 95, 139
Thermal fatigue, 96
Thermal properties, 32, 94–98
  and electrical properties, 94
  ideal and real, 138–142
Thermal shock, 96, 188
Thermionic emission, 36
Thomson effect, 97
Thoriated nickel, 185
Thoriated tungsten, 185
Tilt boundary, 112
Tinplate, 176
Torsion, 87
Transitional elements, 24
Tungsten, 184
Twin band, 113
Twin boundary, 112

Ultimate tensile strength, 82
Undercooling, 139, 141
Unit cells, 59–60
  reference axes, 61
Unit cube, 125, 127

Vacancies, 107, 111, 142
Valency electrons, 26, 28, 36, 43, 48–49, 51, 52, 55, 94, 97, 137
Valency orbital systems, 90
Van der Waals bonds, 30
Van der Waals forces, 30, 38, 44, 66, 121
Vaporisation, 37
Vapour, 31
Vapour phase, 31
Vapour state, 35
Vibration, 14
Vickers hardness system, 77
Vinyl chloride molecule, 29
Volt, 89, 90

Water molecules, 56
Wave, 14
Wave mechanics, 22
Weber, 92
Weldability, 6
Whiskers, 179
White-metal, 181
Wood, 173, 175–176
Work hardening, 136, 157, 158

X-rays, 20

Yield point, 81, 136
Yield strength, 157
  compression, 85
Yield stress, 81, 135
  shearing, 86
Young's Modulus of Elasticity, 80

Zenon, 55
Zirconia, 185
Zone refining, 152